NEUROENOLOGY

NEUROENOLOGY

How the Brain Creates the Taste of Wine

Gordon M. Shepherd

COLUMBIA UNIVERSITY PRESS
NEW YORK

Columbia University Press
Publishers Since 1893
New York   Chichester, West Sussex
cup.columbia.edu

Library of Congress Cataloging-in-Publication Data
Names: Shepherd, Gordon M., 1933– author.
Title: Neuroenology : how the brain creates the taste of wine /
Gordon M. Shepherd.
Description: New York : Columbia University Press, [2017] |
Includes bibliographical references and index.
Identifiers: LCCN 2016021877 (print) | LCCN 2016023079 (e-book) |
ISBN 9780231177009 (cloth : alk. paper) | ISBN 9780231542876 (e-book)
Subjects: LCSH: Wine tasting. | Food—Sensory evaluation.
Classification: LCC TP548.5.A5 S54 2017 (print) | LCC TP548.5.A5 (e-book) |
DDC 641.2/2—dc23
LC record available at https://lccn.loc.gov/2016021877

Columbia University Press books are printed on permanent and durable acid-free
paper.
Printed in the United States of America

c 10 9 8 7 6 5 4 3 2 1

JACKET IMAGE: TK
JACKET DESIGN: Milenda Nan Ok Lee

To Grethe

Contents

CONTENTS

PART III
How Central Brain Systems Create the Pleasure of the Taste of Wine

Preface

This book builds on the principles I presented in *Neurogastronomy: How the Brain Creates Flavor and Why It Matters*. It is to some extent a modern update of Anthelme Brillat-Savarin's classic, *The Physiology of Taste* (1825). There is finally enough experimental and theoretical physiology to begin to answer some of the questions raised in that treatise, and the taste of wine serves as an excellent example.

I have many to thank for assistance in this endeavor. First, my editor Patrick Fitzgerald, aided by my colleague Stuart Firestein, of Columbia University. In *Neurogastronomy*, I devoted much of a chapter to how the brain creates the taste of wine, but it only whetted their appetite for more. Patrick quietly but persistently urged me to expand it into a book. We agreed that it would not be a traditional wine book with a focus on the wine; rather, it would be on an entirely new subject: how the *brain* creates the taste of wine, based on principles that would apply to all wines. As will be seen, the take-home message is: the taste is not in the wine; the taste is created by the brain.

A key role in this book has been played by my wife, Grethe, who over the years has organized visits with many friends in Paris and elsewhere in France. In my interactions with neuroscience colleagues there I became a friend of Jean Didier-Vincent, of the Universities of Bordeaux and Paris, a leading gourmet and wine expert. In 2000, he invited me to participate in "Du Vin" (On Wine), an episode of his radio program on science, through which I met Jean-François Moueix, the owner of Petrus, the world's most expensive wine, and his chief wine taster, the renowned

Jean-Claude Berrouet. Jean-Claude kindly agreed to give me a personal wine tasting to test my theories on how the brain creates the perception of flavor, which stimulated me to begin thinking about writing this book. Notes I made from this special tasting are in the appendix.

One of the principles I have followed in writing *Neurogastronomy* and now *Neuroenology* is to build on my laboratory experiments on the sense of smell. For this I draw on a vast literature in biomedical science, a literature that is revolutionizing our knowledge of the mechanisms underlying normal and abnormal brain function. There is much in this literature that gives insight into the dominant role of smell and the roles of the other senses in creating our perception of flavor, but little of it has yet entered the mainstream of writings on the flavor of food or the flavor of wine. In addition, there is little on the motor control and muscle side of wine tasting, which is crucial to moving the wine through our mouth and the volatiles through our airways.

The challenge in this book is to make these scientific advances interesting and relevant to the experience of wine tasting. I only hope I can impart a little of the sense of excitement that I and my colleagues feel in our scientific studies. All failings and errors in this attempt are entirely mine. Wendolyn Hill has again converted my sketches into elegant illustrations. Pamela Nelson, Irene Pavitt, and Milenda Lee have provided wonderful support in bringing the manuscript through production.

I have benefited from collaborations with and advice from many colleagues. They include flavor experts Dana Small, Ivan de Araujo, and Barry Green at the John B. Pierce Laboratory and Yale University; Gary Beauchamp, Marci Pelchat, and Julie Menella of the Monell Chemical Senses Center in Philadelphia; Terry Acree, Cornell University; Chris Loss, Culinary Institute of America; Harvard anthropologists Richard Wrangham and Dan Lieberman; paleontologist Timothy Rowe of the University of Texas, Austin; Pierre-Marie Lledo, Pasteur Institute; Gordon M. G. Shepherd, Northwestern University; Marina Bentivoglio, University of Verona; Marta Dizy Soto, University of Rioja; and Anna Menini, International School for Advanced Studies, Trieste.

For my education in wine tasting I'm especially grateful, in addition to Jean-Didier and Jean-Claude, at various times over the years to Ann Noble at the University of California, Davis; Sandrine Garbay, Château d'Yquem; Albert Scicluna, Le Rouge et le Verre, Paris; Christian Margot,

Firminich, and Marilisa Allegrini of Allegrini Estates, Verona. Special thanks to Marilisa and to Giorgio, Antonio, and Simonetta Gioco of 12 Apostolis, Verona, for my own Valpolicella wine label. Grethe and I are grateful for a recent invitation from Daniel Baron to join Jean-Claude at the launch of a new Twomey Cellars wine at Charlie Palmer's in New York.

In my current research, I am grateful to my colleagues Michael Hines, Tom Morse, Michele Migliore, Ted Carnevale, Robert McDougal, Perry Miller, Rixin Wang, and Luis Marenco for their continuing stimulation every day. Pasko Rakic and Pietro De Camilli have provided unstinting support as my department chairs. I am indebted to colleague Daeyeol Lee for many stimulating conversations in systems neuroscience. Charles Greer is a longtime colleague and source of much support and wisdom.

Our research is supported by the National Institute for Deafness and Other Communicative Disorders within the National Institutes of Health, with special thanks to Barry Jacobs over the years, and Susan Sullivan. There is much current interest in neuroscience in complete descriptions of brain systems underlying behavior, from molecules to behavior, an effort termed "systems biology." The "human brain flavor system" described in *Neurogastronomy* and this book beautifully illustrates this concept.

NEUROENOLOGY

INTRODUCTION

A New Approach to Wine Tasting

There are many books on *enology*, the field of wine and wine tasting. They describe growing the grapes, producing the wine, marketing the wine under the labels of different vineyards and vintages, and discriminating the tastes of thousands of wines and vintages (for example, Marian Baldy, *The University Wine Course: A Wine Appreciation Text and Self Tutorial*; Désiré Gautier, *Initiation à la degustation des vins*; Ronald Jackson, *Wine Tasting: A Professional Handbook*; Jancis Robinson and Julia Harding, *The Oxford Companion to Wine*; and Michael Schuster, *Essential Winetasting: The Complete Practical Winetasting Course*). Anyone seriously interested in wine tasting needs to have some familiarity with the principles explained in these and many other books.

The Taste of Wine Is in the Brain

This book builds on these authoritative accounts by focusing on a new approach to wine tasting that can be summed up in the phrase: *the taste is not in the wine; the taste is created by the brain of the wine taster*.

How the brain creates the taste—the flavor—of food and drink is a new field, based on the latest research in neuroscience, biology, psychology, and even anthropology. My own field is sensory physiology, with a focus on smell. Recently, I drew together these new findings in *Neurogastronomy: How the Brain Creates Flavor, and Why It Matters*, to explain how the mechanics of food in the mouth and molecules in the

I

airway activate the brain to create the flavors of food. The book has appealed to casual readers as well as many chefs and experts in gastronomy and nutrition. There is now an International Society of Neurogastronomy, reflecting the shared goals of many disciplines and communities to better understand food and flavor and their ramifications throughout society.

This new approach has engaged people interested in applying these principles to wine. At present, no books focus on how wine taste is created by the brain and the biomechanics of the mouth and respiratory system. *Neuroenology: How the Brain Creates the Taste of Wine* aims to expand the field of enology to explain this, whether yours is the brain of the casual wine drinker at home or the brain of an expert professional wine taster. A brief introduction was included in a chapter in *Neurogastronomy* and in a recent article with the same title as this book. Building directly on the principles laid out in *Neurogastronomy*, *Neuroenology* explains how the fluid mechanics of the wine in your mouth and the patterns of your breathing activate your sensory and motor pathways to create the taste of wine and, together with your central brain systems for emotion and memory, generate the whole perception of wine taste. We show that, just as with creating the flavors of food, creating the flavors of wine engages more of the brain than any other human experience.

A Strategy for Neuroenology

Although "neuroenology" is defined as how the brain creates the taste of wine, everything in the brain depends on what is going on with the wine in the mouth, perhaps even more than when we are eating. The movement of a liquid through a physical space in the body is called *fluid biomechanics*. We therefore adopt a strategy of starting a wine book not with the perception of wine but with the fluid biomechanics of the wine through the mouth and the movement of air through the throat and nose. Once you understand these two movements, it will be much easier to see how they activate the brain to create the wine perception.

A second step in this approach is learning that wine tasting depends on structures and muscles that are among the most intricate in the body. Simplified diagrams let you appreciate the amazing ways they function. This will give unique insights into the muscles that move your tongue, the delicate bone in your throat that anchors tongue muscles and swallowing, the way that sniffing the aroma of wine depends on exquisite control of one of the biggest muscles in your body, and the different muscles that you control consciously and unconsciously.

A third step is to explain how much of taste is due to smell—to the molecules released from the wine that are carried to the sense cells in our noses when we breathe out. This is called *retronasal smell* (retro = backward direction; also called *internal smell*) because it goes in the opposite direction from breathing in (*orthonasal smell*; ortho = forward direction). Most wine books recognize the importance of retronasal smell, but there has been very little understanding of how it actually works. Many wine experts on the Internet do not even mention it. We will provide answers to the questions of what exactly retronasal smell is, how it functions, and how it plays the leading role in the taste of wine.

Fourth, we will show how taste itself is an illusion. Most books on wine cover the main sensory systems, starting with the sense of taste, and then describe the other senses and how, together with smell, they create what is called the "taste" of the wine. Although it is called "taste" because it seems to be coming from your mouth where the wine is, this is an illusion because we are unaware of the contribution of retronasal smell. The wine molecules stimulating retronasal smell come from both the mouth and the throat. One of the goals of this book is to explain how this taste illusion is created so that the wine experience can be analyzed and enjoyed more fully.

Fifth, another illusion is that wine tasting is only a sensory experience. The motor activity you use to move the wine around in your mouth while sampling its internal smell is an essential part of creating the wine taste. This makes wine tasting an active sensation. One of the most important motor actions, swallowing, exposes the wine to the greatest possible extent to internal smell. For most people, swallowing simply clears the mouth for the next drink, but we show how important it is for obtaining the most information possible about the quality of a wine.

Finally, this book draws on a lifetime of research, teaching, writing textbooks, and lecturing to diverse audiences on sensory physiology in general and the sense of smell in particular. I especially want to bring together the growing amount of neuroscience research on sensory physiology and flavor as it relates to wine; research that I hope will give exciting new insights into many aspects of the wine experience. The bibliography contains many of these resources.

———————

Creating the taste of wine is a dynamic activity. Stimulation is accompanied by adaptation as the wine taste fades over time; it needs to be kept fresh. Part of this dynamic activity is that, for most of us, wine tasting usually takes place in a social setting, with good company and conversation. Our sensory experience is heavily dependent on our own memories and emotions and those of our companions. All of these factors, communicated by language, mean that wine, like most foods, is subject to influence from our environment. We will show that this influence can be so strong that white wines can be judged to be reds and cheap wines judged to be better than expensive ones. The brain plays these tricks, so it pays to know how it accomplishes them.

While this covers a consensus of research and practical experience in wine tasting, a number of controversies indicate the need for more progress. They concern how our mouth, respiratory tract, and brain actually interact during wine tasting and how we can improve that interaction. We will point them out as we take up motor, sensory, and central systems and indicate in each section the progress that is being made toward solving them.

We tie all this together by breaking wine tasting down into a series of steps, from initially thinking about the wine to the aftermath of swallowing the wine. A unique series of views of the inside of the head and brain will illustrate the steps so that you can see for yourself how the biomechanics of the wine in your mouth and the dynamics of breathing activate the brain systems to create the wine taste.

Finally, as we explain these steps, I will relate them to lessons learned from, among others, a great wine taster, Jean-Claude Berrouet, of Pétrus and Twomey Cellars. Berrouet was a student of another great wine taster, Emile Peynaud, author of *The Taste of Wine: The Art and Science of Wine Appreciation*. We will finish with a diary of my tutorial with Jean-Claude. Casual wine drinkers may be interested in how a top quality wine tasting works, and experts may be interested in how one of the greats organized a tasting for a new convert. You will see that resulted in an unforgettable experience on how empirical principles are applied in practice.

In summary, *Neuroenology* will enlarge your appreciation for how extensively you and your brain are involved in the taste of the wine you drink. I hope you will understand a little better the claim we made at the start: *the taste of wine is not in the wine but in your brain.* Welcome to a fabulous world of wine: inside your head. To paraphrase the Bard, "Oh brave new world, that has such brain cells in it!"

PART I
Fluid Dynamics of Wine Tasting

Most books on wine tasting focus on the sensory sensations because that is what we are most aware of. However, as indicated in the introduction, all of the sensations created by the brain are due to the movement of the wine in our mouth and throat and the movement of the volatile molecules released into the air in our respiratory tract. The relations between them are shown in figure I.1 for an instant in time when the wine drinker has wine in the mouth and is breathing at the same time. These movements generate the activity that stimulates the brain. This diagram will thus serve as a reference for explaining in the rest of the book how the brain creates the taste of wine.

The Fluid Dynamics of Wine and Volatiles

The movement of wine and its volatiles invites study by the principles of fluid dynamics—the flow of liquids and gases in nature. As applied to liquids it is called hydrodynamics and is essential for constructing efficient water supplies and moving oil from refineries. In wine tasting, moving the wine requires propulsion by the muscles of the mouth and throat, involving the activity of many muscles. Although we are usually unaware of them, these muscles are among the most complex in the body. Understanding their principles can therefore add greatly to our experience of wine.

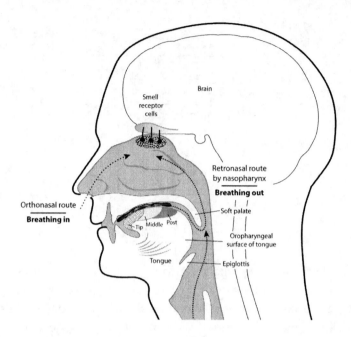

FIGURE I.1

The human head, showing the oral cavity (mouth) and respiratory tract (*shaded*) and the brain above. Wine moves from the tip to middle to posterior sites on the tongue. The airway flow dynamics for sniffing in (orthonasal smell) and breathing out (retronasal smell) are indicated by arrows.

From a broad perspective, these muscles are involved in transporting and processing three types of food. One is *solid food*, such as plants, meat, and seeds, which require active chewing and reduction to a mash before swallowing. Another is *water*, which is usually simply ingested and swallowed. Both of these activities are essential for nutrition and survival. The muscles of the mouth and throat are assumed to have adapted over millions of years of prehuman and human evolution to handle efficiently both food and water. It will therefore be worth our while to understand these basic mechanisms.

The third type of food is flavored fluids; in our case, wine. Wine is a food that is consumed almost entirely for its flavor, with little nutrient value. Wine making developed only in the last several thousand years, too brief a time to have a significant impact on human evolutionary adaptation. When we drink wine, we therefore use the mechanisms for eating that are essential for survival and adapt them for pleasure. We will focus

BOX 1.1
The Main Steps in the Fluid Dynamics of Wine

Taking a sip
Mixing with saliva
Movements of the tongue
Movements of the mouth and jaw
Gating by the soft palate
Gating by the epiglottis
Swallowing

on the principles that have emerged from this research on food in the mouth and show how they can be applied to wine.

In addition to moving the wine we need to understand the transport of volatiles released from the wine into the air in the respiratory tract. As already mentioned, this respiratory flow is essential, first, for sniffing in, called orthonasal smell, which is the aroma of what we are about to eat or drink, and second, for breathing out to carry the aroma from the wine in the mouth and throat to the nose during expiration, to add retronasal smell to the total sensation of flavor from the food in the mouth. We will consider the muscles involved in producing both of these contributions to the flavor of wine.

In this new approach to understanding wine tasting, we have indicated the overall sequence of events that takes place. This sequence can be broken down into steps. Some of the steps occur one after the other, while others overlap. Some involve the flow of the wine, and others involve the flow of air. The steps involved in the movement of the wine are summarized in box I.1.

The Composition of the Air

The two kinds of air in orthonasal and retronasal olfaction have interesting similarities and differences. Orthonasal air comes exclusively from the environment and consists of the gases nitrogen, oxygen, and carbon dioxide, together with evaporated water varying with temperature and humidity. It includes various trace compounds, depending on the local environment and levels of pollution, as well as particulate matter, such as smoke, pollen, dust, and lint. Finally, it contains many different types of volatile molecules, depending on our activities, that come from industrial exposure to open nature to mealtimes when we are sampling the scents of food and drink. Although not often noticed, at high levels these external conditions can distort flavor perception, as in smokers and in polluted environments.

Retronasal air, in contrast, reflects the characteristics of the expiring air that contains contributions from inside the body after it is filtered by the upper airway. In addition to nitrogen, oxygen, and carbon dioxide, it has high humidity due to moisturizing by the airway mucus membranes and a warm temperature due to the body's temperature. Among its trace elements are the volatile compounds that have been breathed in, and those called *endogenous volatile organic compounds* (VOCs) that come from metabolic processes in the body. Although not usually considered significant, we will mention three—acetone, isoprene, and alkanes—that may seem surprising to find in the body and are of possible relevance to wine tasting.

Acetone is actually an industrial solvent. It is produced by glucose metabolism and circulates in the blood in trace amounts. It is raised in high-fat, low-carbohydrate diets. It is especially high in diabetes and alcoholism and may be raised during pregnancy and nursing.

Isoprene is slightly fragrant. It is produced by many trees and plants and therefore present in inspired air. (As a polymer, it forms the basis of rubber.) In the body, cholesterol metabolism produces the molecule; it is a subunit of vitamin A and involved in photoreception in the retina. It is present in exhaled air; humans smell slightly of isoprene.

Among the *alkanes*, the simplest is methane, produced by intestinal digestion and a contributor to global warming through millions of domestic animals.

Finally, it may be noted that the ingestion of wine itself puts an increasing load of *alcohol* in the blood, which lends a characteristic sweet fragrance to the exhaled breath.

These substances enter retronasal smell via several routes. One is by diffusing from the lung capillaries into the air in the tiny pockets called *alveoli* that make up the lung. Another is by seepage into the airway within the lung, trachea, and pharynx. And a third is from the wine in the mouth, as we shall see.

In summary, inhaled air and exhaled air are similar in their basic gases but have distinct combinations of volatile molecules from the environment and from the body. For neuroenologists, two questions seem particularly interesting. First, can these differences in volatile molecules contribute to differences in the wine aromas of orthonasal compared with retronasal smell? Second, can they contribute to the differences in sensitivity to smell between the two routes? These questions deserve further study.

The pattern in normal people can vary under different conditions of age and health. For example, a study of centenarians found differences in VOCs compared with younger populations. And many studies have documented the effects of different diseases. Kidney failure; cancer; and neurodegenerative, neurological, and psychiatric disorders cause metabolic and related changes: the blood can convey these metabolic effects to the lung and out into the exhaled air. It is possible that these changes can be detected as early signs of these disorders and could impinge on the aromas sensed in wine tasting. Evidence that isoprene and acetone can vary in the exhaled breath under different emotional states suggests that this could be an area of active research in the future.

The Airflow Dynamics of Wine Volatiles

Airflow dynamics engineers study physical devices such as airplane wings and heating systems in order to design optimal airflow patterns. The flow of volatiles in wine tasting is fast enough for these principles of airflow dynamics to be applicable. Scientists have already begun to apply this analysis to airflows in the nasal cavity during orthonasal smell. Here we expand this approach to embrace the flows not only in the nasal cavity but also in the nasopharynx and oropharynx (throat), during both orthonasal and retronasal smell. This is a step toward creating a new field of *wine volatile flow dynamics*, which will help to move the mouth and throat from our general ignorance about their structure and function into the mainstream of research on the human body.

BOX I.2
The Main Steps in Airflow Dynamics of Wine Volatiles

Sniffing in: orthonasal air currents in the nose
Breathing out: retronasal air currents in the nose
Breathing out: movement of volatiles from the mouth
Breathing out: movement of retronasal air currents from the pharynx
Breathing stopped: during the swallow
Breathing out: the "flavor burst" and "finish" after swallowing

Engineers develop flow dynamics on a highly quantitative basis, but we will use it mainly on a qualitative basis, to explain the principles involved (box I.2). This will reveal a much richer understanding of how the perceptions of wine tasting arise when we take up the steps of sensory processing.

The stages of moving the wine through the mouth end with the automatic act of swallowing, which forces the wine from the mouth into the throat and through the throat into the esophagus. This leaves a coating of wine on the mouth and throat, whose volatiles are swept up by the respiratory system and transported to the nasal cavity.

In this part, on muscles and flows, we begin in chapters 1 and 2 with the fluid dynamics of the wine in the mouth as a basis for stimulating the sensations of mouthfeel and taste as well as producing the volatiles. Chapters 3 and 4 consider the fluid dynamics of the volatiles in the respiratory tract that produce orthonasal and retronasal smell. Chapter 5 describes the swallow and its aftermath, which make a large contribution to retronasal smell and the wine aroma. This is the final step in the wine-tasting experience.

Given the many factors involved in muscle actions and the many routes for the dynamic flow of fluid and volatiles, it is not surprising that unsolved problems and controversies exist. Box I.3 summarizes a few of them. We will find answers and raise new questions in the following chapters.

BOX I.3
Controversies About the Fluid Dynamics of Wine Tasting

How does the fluid dynamics of wine tasting compare with the fluid dynamics of food and water in the mouth?

What are the stages of fluid dynamics in the mouth, from the initial sip to swallowing, and how are they related to the sensory experience?

To what extent are the differences in evaluating wines due to differences in fluid mechanics or differences in individual brains?

Wine volatiles arise from the mouth and, after swallowing, from the mouth and throat. Which makes the greater contribution to the taste of wine?

CHAPTER ONE

Sip and Saliva

Fluid dynamics of wine in the mouth, we have seen, means the factors that move the wine from the initial sip through the mouth to the throat and into the esophagus and stomach. This entire sequence, from sip to swallow, is called *deglutition* (literally, the transport of contents out of the mouth). The key questions for neuroenology are, first, how the fluid stimulates the mouthfeel and taste systems in the mouth and, second, how it gives rise to volatiles that contribute to the wine taste through retronasal smell.

Modern studies of deglutition started in the 1950s and focused mostly on eating food. The early studies were based on hazy fluoroscopic images of human subjects with radiodense food or water in their mouths, giving the impression that the mouth creates a closed cavity in which a seal between the back of the tongue and the soft palate (*vellum*) closes the mouth's contents off from the flow of air in the pharynx. If this were true, no connection would exist between the mouth and the nasal cavity, and retronasal smell could not occur while wine is in the mouth. This implies that retronasal smell can contribute to flavor only after swallowing, leaving volatiles on the membranes of the throat where expiration can carry them to the nasal cavity. Over the years, as the images have improved, it has become clearer that even with food and water—and by implication, wine—the back of the mouth can be open to the pharynx to a significant extent, enabling volatiles in the mouth to stimulate retronasal smell.

These mechanisms of dynamic flow lie at the heart of the wine-tasting experience. It will repay our effort to consider them in detail. We start by expanding somewhat the series of steps, introduced in the introduction, through which wine moves in the mouth (box 1.1).

We will discuss each of these steps in order to reconstruct the mechanics underlying wine tasting. In this chapter, we consider the sip followed by its mixing with saliva.

BOX 1.1
Fluid Dynamics of Wine in the Mouth

Collection phase: the sip

The sip is collected in the anterior mouth while the back of the tongue is raised to the soft palate to hold the wine in the mouth.

Front of the mouth phase: secretion of saliva

The wine combines with saliva, which dilutes and subjects it to enzymatic actions.

Front of the mouth phase: retronasal smell

The wine is held in the mouth by the tip of the tongue pressed against the gums under the front teeth and the back of the tongue pressed against the soft palate.

Middle mixing phase: tongue, cheeks, jaw

The tongue agitates the wine by curling and swishing it against the cheeks. The jaw moves up and down to agitate the wine by "mastication."

Posterior tongue phase

The tongue rises toward the hard palate to move the wine toward the back of the mouth.

Palatal block phase

As the wine approaches the back of the mouth, the soft palate rises to block the nasopharynx to prevent the wine going into the nose while the epiglottis closes to prevent the wine going into the trachea.

Pharyngeal (swallowing) reflex phase

Pressing the tongue against the hard palate pushes the wine into the pharynx. All mouth movement and respiratory movement stops. The wine passes into the esophagus and stomach.

Postswallowing phase

The tongue relaxes, the nasopharynx and epiglottis reopen, and breathing resumes.

Adapted from a consensus of many studies of the movement of food and liquids.

The Sip

Obviously, the first step in the fluid mechanics of wine tasting is taking the sip, which places the wine in your mouth.

The sip is critical to wine tasting for several reasons:

First, it is a fateful step in the perception of the wine taste because the feeling of the wine in the mouth gives the illusion that all the subsequent sensations of the flavor of the wine come from the mouth. We will discuss how this sensory stimulation happens in chapter 10.

Second, the amount of wine in the sip can have a critical effect on the perception of the wine. In many restaurants, the waiter often pours the glass full, as a generous sign of welcome, but this has several disadvantages: it does not leave enough room in the glass for the bouquet to form; it does not leave enough room to swirl the wine in the glass; and it encourages taking large mouthfuls, which overwhelm the taste buds (and lead more quickly to inebriation).

For the serious wine taster, the answer is: take the amount that will best enhance the taste of the wine. This means not too little to directly stimulate all the touch and taste sensory receptors in the mouth and indirectly stimulate the nose by means of the volatiles released, and not so much that it cannot be adequately controlled in the mouth, which, as we shall see, limits the contribution of the wine in the mouth to retronasal smell. Experts, such as Ronald Jackson in *Wine Tasting: A Professional*

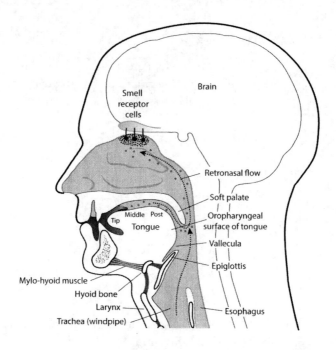

FIGURE 1.1
The sip places wine on the anterior tongue, which begins the illusion that all wine taste comes from the mouth, even though the volatiles begin to stimulate the olfactory receptors by retronasal smell.

Handbook, have their optimal amounts, which usually vary from 6 to 10 milliliters; for Americans, that is roughly equivalent to 1 to 3 teaspoons, or up to 1 tablespoon (table 1.1). If you are strictly an amateur with wine, it can be interesting to test yourself by taking sips of water from a teaspoon or a tablespoon. For most amateurs, 1 tablespoon will probably seem like enough in your mouth.

Third, a small sip is significant for another reason: just after entering the front of the mouth, the wine is in its purest form, still equivalent to the wine in the glass before it starts to mix with saliva. Keeping the small amount of wine in the front of the mouth, you can lower the back of your tongue to let the volatiles reach the back of your mouth and then release into the air on expiration, thereby immediately bringing the contribution of retronasal smell from the still-pure wine into the wine flavor.

TABLE 1.1
How Much Is a Sip?

1 milliliter = 0.04 ounce
1 milliliter = .004 cup
1 milliliter = 1 cubic centimeter
1 tablespoon = 14 milliliters
1 teaspoon = 5 milliliters
1 tablespoon = 0.5 ounce

How can you sense the wine aroma by retronasal smell even though it is still in the front of the mouth? This will be explained in chapter 4.

Wine + Saliva

The next step of wine in the mouth involves its mixing with saliva.

Although the initial sensing of the pure wine benefits from a limited sip, the subsequent perception benefits from a larger sip. You can test for this by sipping some wines to see whether you can move the wine in your mouth in a way that adequately exposes it to all the surfaces of the tongue and cheeks without overstimulating them, as well as get a strong impression of the added flavors due to retronasal smell when you breathe out. You may discover that you have been understimulating or over-stimulating yourself with the wine and be able to determine the amount that gives you the most rewarding experience.

Even before the wine enters your mouth, the salivary glands begin to secrete saliva. This is because when we see wine or think about drinking wine, our brains activate a central control center called the *hypothalamus*, which in turn activates the *autonomic nervous system*. These nerves regulate our internal organs, heart rate, respiration, digestive tract, sexual arousal, and all the other vital functions, without our having to think about them. This system has two types of nerves: the *sympathetic nervous system*, which responds relatively quickly to change, such as increasing the heart rate when we are excited, and the *parasympathetic nervous system*, which responds relatively more slowly, such as slowing the heart rate when we are resting. However, as in the case of the salivary glands, sometimes they act together, receiving both types of nerves to exert coordinated control. So whether your mood is excited or subdued can affect how much saliva you produce and its composition as you drink the wine.

Saliva is produced and secreted throughout the day and night. The amount varies widely among individuals, from around 1 to 3 tablespoons an hour (1 to 3 pints a day), almost all of which you continually swallow. The production is highest in the late afternoon and lowest at night. Of course, this rate increases briefly when we see or think about food or wine and when we consume it. This means that when we describe the taste of wine we are actually describing the taste of "wine + saliva." Since the amount of saliva varies substantially among individuals, this can be a significant variable in wine tasting. And the variation in saliva during the day means that a wine can taste different at lunch and at dinner.

Glands that make two kinds of secretions produce the saliva (table 1.2). One secretion is watery and is called *serous* fluid; the other is thicker and is called *mucus*. The glands are situated around the mouth. Most saliva is secreted from the *submandibular* (below the jaw) glands; you can feel them as two small balls just under your jaw. They secrete a mixture of serous fluid and mucus. The large *parotid* (from the Greek *para* [near] and *otid* [ear]) glands are situated against the back of the jaw in front of the ear; they secrete a serous fluid that combines with the wine to bathe the taste receptors on the tongue and enhance their stimulation. The watery fluid also helps smooth the way during swallowing. The saliva contains enzymes that begin to digest starches. A small amount of mucus saliva is secreted by the *sublingual* (under the tongue) glands.

TABLE 1.2
Salivary Glands and Saliva

Gland	Type of saliva	Proportion of saliva (percent)
Submandibular	Mucus and watery	60
Parotid	Watery	20
Sublingual	Mucus	5
von Ebner's	Watery	5
Gland cells	Mucus	10

BOX 1.2
Functions of Saliva

Smooths the mouth surface
Keeps taste buds moist
Dissolves taste substances
Digests taste substances
Opposes astringency in wines
Warms the wine to enhance volatile release

The von Ebner's glands, situated toward the back of the tongue, are a special type of glands; they secrete a serous fluid that bathes special taste receptors in this area. The secretions contain enzymes that begin the digestion of lipids (fats). Although wine does not contain much lipid, food eaten with wine often does. In addition to these distinct glands, hundreds of tiny *clusters of gland cells* within the membranes of the mouth secrete mostly mucus that lubricates the surface of the mouth.

The saliva has important functions in relation to wine tasting. First, it keeps the membranes lining the mouth moist so the wine can move smoothly over the surface. The smoothness of movement depends on the balance between the serous and mucus parts of the saliva. We will see that the astringency of wines can significantly alter this smoothness. Second, the saliva keeps the taste buds moist, enhancing the ability of taste molecules and ions (tiny electric charges) to stimulate the taste cells. Third, adding warm saliva to the ingested wine, especially cool white wine, enhances the release of volatiles to retronasal smell. Finally, the salivary secretions contain enzymes that help to break down and predigest components in the wine, especially carbohydrates. Box 1.2 summarizes these important effects of saliva in wine tasting. Most of these have the same effect on food eaten with the wine.

These actions of the saliva change the character of the wine from the time it enters the front of the mouth to the time it exits at the back. A wine taster therefore experiences the wine differently as it moves through the mouth. This passage is controlled by the master conductor of the oral orchestra, the tongue.

CHAPTER TWO

The Tongue
Moving the Wine

As soon as it is sipped, the wine needs to be moved in a controlled manner throughout the mouth to stimulate maximally all the senses, facilitate the delivery of volatiles to the back of the mouth, and participate in swallowing. These biomechanical steps are critical to the wine taste, and the tongue is the master tool for doing it.

The Tongue Goes to Work

As shown in figure I.1 and in further detail in figure 2.1, experts have identified different parts of the tongue that are involved in moving food and liquids from their point of ingestion to their point of exit. Each part is associated with a step in the transport of the wine. The first step is receiving the sip just behind the gums, where it is immediately tested by the tip of the tongue for its tactile "mouthsense" and by taste buds for its taste. In this initial stage, as we have seen, the wine is least dissolved in saliva. Importantly, it is also furthest from the back of the mouth so that the soft palate (*velum*) can be raised without wine being aspirated into the trachea or the nasal cavity. The volatiles released by the wine can thus be entrained by expiring air in the pharynx and carried to the nasal cavity for retronasal smell while the wine is still in the mouth. This is also the opportune time to aerate the wine (chapter 3).

The tongue can then form a shallow depression on its dorsal surface to transport the wine and saliva backward. Middle and posterior tongue

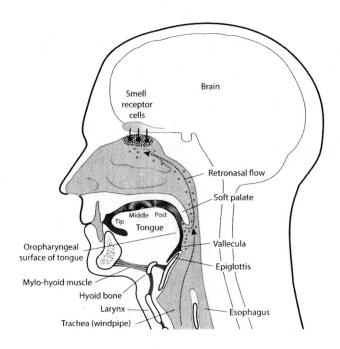

FIGURE 2.1

View of the oral cavity (mouth) and oropharynx (throat), showing the wine moving down the back of the tongue to the valleculae, where the volatiles are picked up by retronasal airflow.

sites have been identified from experiments on eating food. At these sites, vigorous chewing movements of the jaw agitate the wine to promote release of the volatiles. At the posterior site, the wine begins to leak over and down the back of the tongue, ending in the two pockets next to the epiglottis called the *valleculae* (little valleys). This leaked wine, of course, contributes volatiles to retronasal smell.

During this transport, the contents of the mouth are also being thoroughly mixed with saliva and subjected to salivary enzymes. In the case of food, mastication (chewing) is also occurring, reducing the food to a finer and finer mash.

In addition to transport, the contents of food are also being processed en route. In the case of liquids such as wine, there is also what one can call *agitation*, involving swishing the tongue and chewing to move the wine and generate maximum sensory stimulation, including pushing the wine everywhere against the walls of the mouth cavity—the cheeks on the side, the hard palate above, the gums in front, and the taste buds on the tongue.

In order for the tongue to move food and wine about in the mouth in this way, its surface needs to be rough. You can feel this roughness yourself by moving your tongue against your hard palate. This roughness is due to many tiny outcroppings from the tongue's surface called *papillae*. Most of the roughness is due to filiform (hairlike) papillae, which have up to a dozen or so hairlike extensions covered by a hard layer of keratinized cells. These do not have any taste buds in them, serving only a mechanical role in the movement function of the tongue. Other types of papillae are shallower outcroppings containing taste buds, which will be described in chapter 11.

The Muscles of the Tongue

When you stick out your tongue and examine it in a mirror, it seems to be one big muscle, but in fact it is composed of many muscles. As shown in figure 2.2, four *extrinsic* (externally attached) muscles are anchored to bones that form the floor of the mouth and the temporomandibular joint, the joint between your temporal bone on the side of your skull and the jawbone. Depending on where they are anchored, these muscles have specific and sometimes overlapping functions that move the tongue so that it protrudes, depresses, elevates, retracts, and moves from side to side. Most of these functions are needed in all animals, such as retracting and raising the back of the tongue during swallowing. The tongue also has four *intrinsic* muscles embedded within it. These paired muscles extend the length of the tongue on each side and enable its refined placement and shaping. These movements are necessary for the manipulation of food and liquid in the mouth during eating and drinking.

Given the complex series of movements we have described for manipulating the wine, one gets a better appreciation for how much training is needed for a wine taster to graduate from novice to professional. We can count up the number of muscles involved in moving wine within the mouth. In addition to the four extrinsic muscles depicted in figure 2.2 are the four intrinsic muscles. There are also the muscles of the cheeks and lips and five thin muscles that control the soft palate. Training this

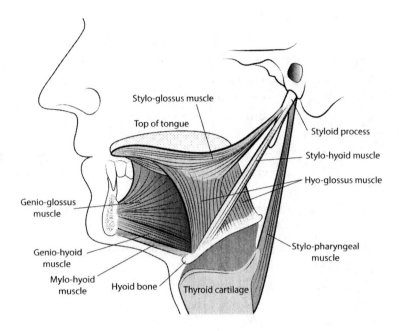

FIGURE 2.2

Large movements of the tongue are controlled by muscles attached to the floor of the mouth and at the temporomandibular joint in the upper right. (Adapted from H. Gray, *Anatomy of the Human Body*, 26th ed., ed. C. M. Goss [Philadelphia: Lea & Febiger, 1954])

set of tongue and mouth muscles to become a professional wine taster seems as demanding as training the muscles of the arm and body to become a professional tennis player! In the appendix I describe how Jean-Claude Berrouet demonstrated his lifetime of training of these muscles during our wine tasting.

The tongue has other functions beyond manipulating food and fluids. It is a prime example of a *multifunction* organ. One of the dramatic developments in human evolution is the adaptation of the tongue for forming the phonemes of speech. It might be supposed that speech requires more refined tongue movements than eating. A multidisciplinary team of long-time investigators of tongue movements in humans during eating, physiologists Karen Hiiemae of Syracuse University and James Palmer of Johns Hopkins University, together with anthropologist Daniel Lieberman at Harvard University, and their colleagues, tested this theory. Surprisingly, they found that the tongue movements involved in eating are more complex and refined than the tongue movements used for speech. The same likely applies to wine, which leads to an interesting paradox: the tongue movements involved in complete and systematic wine tasting are likely to be more complex than those in the speech we use to communicate our perception of the wine we are tasting!

The key sensory sites to which the wine is delivered are, of course, the taste buds on the surface of the tongue and pharynx (chapters 11 and 12). This means that the tongue also contains a complicated sensory organ. We say therefore that wine taste is an *active sense*. If you feel something placed on your fingers, that is *passive touch*, but if you reach out to hold it and manipulate it with your fingers, you will perceive a much more vivid *active touch*, which produces a lot more information. You can test this yourself with any small object. Wine taste is a supremely active sense.

In summary, training to be an expert in tasting wine depends to a great extent on training the tongue to deliver the wine to the wine-sensing organs of the mouth, together with delivering the volatiles to the internal nose (chapter 3).

Middle Mixing Phase: Mouth, Cheeks, and Jaw

We have seen that in addition to the tongue, the mouth, cheeks, and jaw are also involved in active wine tasting. Konrad Burdach and Robert Doty first analyzed these movements in a classic experiment in 1987 at the University of Rochester. It is still instructive to learn from them.

They noticed that food flavors seemed to be more intense during chewing and swallowing and set out to analyze this more closely. Solutions of distinctive flavors of imitation extracts of rum and orange were used. Subjects were asked to report the intensity of flavor under several conditions: no mouth movements, vigorous chewing movements, and after spitting out the solution or after swallowing. The results showed that the flavor intensity was weak without movements and intense after movements, spitting, and swallowing. The authors concluded: "The present study suggests that retronasal odor perception is a highly dynamic process, and that retronasal movement of molecules to the nasal epithelium is likely dependent on air currents induced by active alteration of the musculature of the mouth and pharynx."

The authors found that "mouth movements seem to be a major factor in determining the intensity of the retronasally perceived stimuli." They noted a nice symmetry: vigorous sniffing is an active process underlying orthonasal smell, and vigorous movements of the tongue, mouth, and pharynx are an active process underlying retronasal smell. All of these principles obtained by scientific study were demonstrated in action by Jean-Claude Berrouet in his vigorous "chewing" of the wine described in the appendix.

Cheek Muscles and Retronasal Smell

The mouth provides the boundaries of the arena in which the wine is moved, but these are flexible boundaries. The cheeks move in and out, the jaw moves up and down, and the tongue pushes the wine against the sides of the mouth where the gland cell clusters are found. The tongue also pushes against the roof of the hard palate. Experiments have shown that this actually increases the release of volatiles from the wine.

The main cheek muscle is called the *buccinator* muscle. It is a broad, thin muscle, anchored at the upper and lower parts of the back of the jaw and spreading out across the cheek to attach to the angle of the mouth. When you press a finger on your cheek, you are feeling the buccinator muscle. It works together with the tongue muscles and the muscle that encircles the mouth to keep food between the upper and lower teeth and keep wine moving into all corners of the mouth.

Chewing the Wine

Movements of the jaw are important for "chewing" the wine. This seems counterintuitive: the wine is a liquid, not a solid. However, the wine's astringency and alcohol content give the impression of it having sub-stance, another illusion associated with flavor, and we move our jaw to chew it. This is important in agitating the wine to help release the volatiles.

It also gradually moves the wine on the tongue back toward the pharynx, where it leaks down into the valleculae to be picked up by expiring air and carried to the nasal cavity to stimulate retronasal smell, as noted previously.

Chewing, in fact, is an inherent activity in response to something in the mouth. This vital function is controlled by a group of cells called the *mastication central pattern generator* (mCPG). It has been studied by Arlette Kolta and James Lund at the School of Dentistry at McGill University in Montreal. The mCPG is found in the brainstem among the other vital central pattern generators. Since sensory feedback from the jaw muscles closely control it, it is located within the trigeminal (fifth cranial nerve) sensory nucleus. Like other CPGs, it consists of a core of neurons that have natural bursting properties and are turned on and off by interactions with inhibitory interneurons. The impulse *outputs* go to the muscles that open and close the jaws. The impulse *inputs* from the jaw come from receptors in those muscles that signal the amount of alternating contraction and relaxation, joint receptors at the temporomandibular joint, and even sensory receptors that surround the teeth and signal the amount of impact on the teeth. Another important set of inputs comes from higher up in the brain; these impulses originate in the jaw area of the motor cortex and descend carrying voluntary messages to initiate, control the force of, and terminate chewing. It is through this higher control that learning takes place to modulate chewing to extract the optimal amount of information about the wine being processed in the mouth.

Preswallow Phase

Three muscle groups are essential for finishing eating and preparing to swallow foods and liquids, including wine.

The first muscle group is the *posterior tongue*. As middle mixing begins to end and the tongue begins to move toward a swallow, the back of the tongue starts to rise to continue to hold the wine in. Increasing amounts spread over the back of the tongue and slide down into the valleculae, where exhaling air takes them up and increasingly stimulates retronasal smell.

The second group is the soft palate, which hangs over the back of the mouth, ending in a final small lollipop of tissue called the *uvula* (you can see it in the mirror when you open your mouth). We have noted the five small, thin muscles that control the movement of the soft palate. These movements pull the soft palate down and against the back of the tongue to hold the wine in the mouth and lift it up against the back wall of the nasopharynx to close it off and keep food and drink from going up into the nose when we swallow.

These movements of the soft palate can be controlled voluntarily to some extent. This is important during wine tasting because the soft palate can allow the volatiles arising from the wine to be entrained at the back of the mouth by expiration so they can stimulate retronasal smell while the wine is still in the mouth.

The third muscle group is related to the epiglottis; these muscles and their functions are discussed in chapter 4. As will be explained in chapter 5, with the soft palate blocking the nasopharynx and the epiglottis blocking the airway, swallowing can then proceed. But first, we need to discuss respiration and its importance to the wine-tasting experience.

CHAPTER THREE

Respiration and Wine Aromas

Respiration, like the beating of the heart, is one of the vital functions essential for life. Unlike the others, respiration is special because it is intimately involved in the sensory experience of food and wine. The breathing rhythm is automatically generated in the brainstem, though it can be modified by voluntary will. It consists, obviously, of a continuous sequence of breathing in (inspiration, inhalation) and breathing out (expiration, exhalation).

Wine drinkers make selective use of that sequence. When we want to sense the bouquet of the wine before we drink it, we hold the glass near and breathe in. The perception of the aroma is entirely conscious. When we sense the taste of the wine in our mouths, however, we are breathing out, but we think the taste is all coming from the mouth: the smell that contributes to the taste is entirely unconscious. Since this is one of the new frontiers in wine tasting, we need to start with a better understanding of the breathing mechanism.

Our Breathing Apparatus

The main respiratory muscle in the human is the diaphragm, the large, flat muscle sheet just below the lungs that is anchored to the lower ribs and upper abdomen (figure 3.1). It is shaped like a dome so that when it contracts, it pulls the diaphragm downward, creating suction in the chest that expands the lungs to draw air in as we inhale; on the other

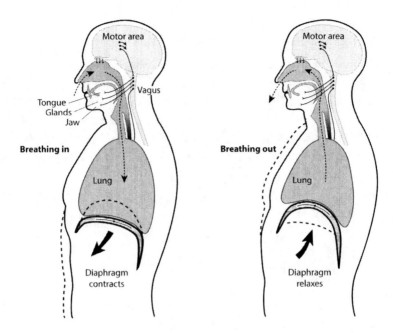

FIGURE 3.1

Movement of air through the nose is controlled by movements of the chest and diaphragm during breathing in and breathing out.

hand, when it relaxes, it pushes upward against the chest to shrink the lungs and push air out as we exhale. The diaphragm receives its innervation from the vagus nerve arising in the brainstem, which in turn is under voluntary control by the motor cortex in the brain, as indicated in the diagram.

During normal breathing, this sequence occurs without us thinking about it. However, rapid inhalation, as in conscious sniffing, brings in the contraction of the muscles between the ribs (intercostals), and rapid exhalation brings in the muscles of the abdominal wall, helping to push air out. Thus, when we are sniffing the bouquet of a wine we think we are sniffing with our noses when we are actually sniffing through our motor control of the diaphragm (see the diagram). It is surprising that one of the largest muscles in the body can be trained to carry out the meticulous control of this delicate movement, aided by various muscle movements and muscle groups that can switch between unconscious and conscious operation. In this way, respiration can be trained to enhance the wine-tasting experience.

The main function of the lungs is to exchange the oxygen in the air we breathe in with the carbon dioxide in the blood produced by the body's metabolism. This has an interesting consequence that is seldom noted in wine tasting: the air we sniff in to test the bouquet of the wine before drinking is high in oxygen but low in carbon dioxide, while the air we breathe out to taste the wine through retronasal smell is low in oxygen but high in carbon dioxide. These and other differences in the composition of the volatiles in orthonasal versus retronasal smell have been summarized in chapter 1.

The Respiratory Central Pattern Generator

Normal breathing is generated by the *respiratory central pattern generator* (rCPG). The cells that make up the rCPG are in the brainstem, like other CPGs involved in vital functions. The cells are not found in one small cluster but are spread throughout much of the brainstem. This perhaps reflects how complex and vital their function is, as breathing must be able to be modulated in many ways (including by wine tasting!), and the output must go to many kinds of muscles.

Although the rCPG has been studied for many years, it is so complex that scientists only have an outline of how it functions. Nonetheless, it is clear that like other CPGs it consists of three cell populations. The first is the central core of cells that have *intrinsic impulse bursting* activity modulated by interneurons that control the pattern of output. The second includes *impulse outputs* that go to premotor cells that then connect to the respiratory muscles. The most important among these is the diaphragm. Other outputs go to the muscles of the ribs and upper body, as well as the neck and face. The third is made up of a wide range of *impulse inputs* that provide feedback from the muscles to precisely match their activity to the particular task at hand—in this case, the different uses of orthonasal and retronasal flows during wine tasting.

We will discuss the orthonasal and retronasal airflows throughout this book. Here, we note that the rCPG can be modulated by voluntary control. You may observe online several videos of expert wine tasters sniffing a wine to identify its varietal and region of origin, often with amazing accuracy. The tasters perform long sequences of precise sniffs while concentrating intensely on evaluating the wine aroma by orthonasal smell. After ingesting the wine, slow expiration is equally important, especially following swallowing, to evaluate the aroma burst and the finish due to retronasal smell.

The Sniff

Even before taking a first sip of wine, the taster first encounters the wine with the initial sniff of its aroma. It seems a rather trivial maneuver, often the subject of flip remarks (". . . nothing to sniff at . . ."). However, it is in fact a critical element of smell. Anyone with a dog as a pet, guide, or guard knows that a dog on a track is completely obsessed, with an exquisite control of its sniffing muscles, from snout to lungs. The same applies to an expert wine taster on the track of the aroma from a wine in a glass. Sniffing is critical during the first encounter with this aroma, and repeated sniffing of the wine in the glass is also critical as the taster refines his or her analysis of the aroma to arrive at a judgment.

Sniffing may be defined as the conscious effort of inhaling repeated forceful intakes of air through the nose. It is a motor adaptation for increasing the detection and discrimination of weak odors in the environment. Perhaps our most common belief about this simple act is that rapid and forceful sniffing increases odor detectability. This belief was tested carefully by Jonathan Beauchamp and a team led by Andrea Buettner from three institutions in Germany: the Department of Sensory Analytics, Fraunhofer Institute for Process Engineering and Packaging in Freising; the Department of Chemistry and Pharmacy, Emil Fischer Center, Friedrich-Alexander-University in Erlangen-Nürnberg; and the Smell and Taste Clinic, Department of Otorhinolaryngology in Dresden. I mention these institutions to indicate the tremendous expertise across a broad range of disciplines that is being marshalled to elucidate smell and taste in humans—the kind of scientific expertise that neuroenology needs to build on.

Beauchamp and his colleagues tested subjects for their subjective ratings of the intensity of the smell of diacetyl (butanedione) under three sniffing conditions: normal, rapid, and forced. In parallel, they measured the concentration within the nasal cavity (at the internal nares and the olfactory cleft, the narrow opening just under the olfactory receptor sheet) under the same three conditions. (Diacetyl, a product of grape fermentation, has a buttery or butterscotch taste and smell. It also adds to the smoothness of the mouthfeel. It is present in low concentrations in some wines, such as chardonnay.)

The results were surprising and counterintuitive (always the best result of a scientific experiment!). The subjects rated the concentration highest with *normal sniffing*, in accord with the finding that the highest odor concentrations in the nose occurred with normal sniffing and the lowest with strong forced sniffing.

These results appear to reflect the fact that diacetyl is a highly soluble, high-vapor-pressure molecule that is quickly absorbed in the mucus where it stimulates the olfactory receptor neurons (chapter 7). By contrast, slow, long-lasting inhalations enhance the ability to detect a low solubility, low-vapor-pressure molecule. Rapid, shallow sniffs may be the best compromise, building up the concentration of both high- and low-solubility odor molecules in the nasal cavity, to access best the olfactory receptor sheet. Sniffing strategy is therefore a critical element in wine tasting.

Another way of explaining this is that odor molecule types differ in their ease of what is called *sorption* into the mucus. High-sorption molecules produce higher olfactory responses at high sniff velocities, and low-sorption molecules produce higher olfactory responses at lower sniff velocities. You can understand this if, at the extreme, a high-sorption molecule instantly enters the mucus, so the faster the velocity, the more uptake; on the other hand, a very low-sorption molecule might have to sit on the mucus for a long time before it is taken up. The tradeoffs reflect different attributes of the flow dynamics of the volatiles. Since wines contain complex mixtures of molecules with different sorption properties, a practiced wine taster may have to explore several different sniff rates to bring out the maximal responses of the different components.

Other considerations related to sniffing include the following: If odors are noxious, sniffing will increase their intake into the lung. Rapid but shallow sniffing enables exposure of the olfactory receptor cells to volatiles in the air without inhaling deeply into the lungs.

Another variable in the motor apparatus of human sniffing is the *nasal cycle*, in which the membranes on the turbinates in the two sides of the nasal cavity alternate in swelling over a roughly 90-minute period. Odor detection is obviously best through the more open nostril with the highest airflow rate.

For optimal wine tasting, the muscles involved in sniffing, from the diaphragm in the abdomen up to the smallest muscles of the nostrils, must be carefully controlled for rate, duration, depth, flaring of the nostrils, and the like. Being a motor act, sniffing is part of the vigorous movements that make flavor an active sense; this applies to the nose as well as to the mouth. As noted, a professional wine taster controls sniffing like a professional tennis player controls ground strokes.

Orthonasal Airflow Patterns in the Nose

As the sniffed air enters the nose, fundamental questions arise regarding how the air reaches the olfactory receptors.

The physiologist Max Mozell at Syracuse University is a pioneer in studying airflow in the nose. He began in the 1950s to test the hypothesis that odors are encoded by spatial activity patterns in the sheet of smell receptor cells. When he found gradients of sensory activity across the sheet, he wondered if they could be due to differences in sensitivity in the receptor cell population. But, could they also be due to different rates of absorption into the mucus layer overlying the receptor cells? The model he thought of was a chromatograph, the basic instrument used in biochemical laboratories around the world to analyze the chemical composition of compounds. The instrument produces the biochemical components of a mixture as a gradient in a readout. By labeling different odors, he showed that there was a gradient of "sorption" across the receptor sheet that was steep for some odor molecules and gradual for others, depending on their fat and water solubility. Because of the resemblance, he called this the "chromatographic hypothesis" of odor coding.

The early experiments were performed on the simple sheets of cells in frog noses. To test the hypothesis, he led a group of collaborators to analyze the flow currents in the human nose. Since the 1980s, studies of

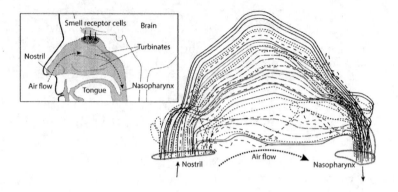

FIGURE 3.2
Orthonasal airflow patterns within a model of the human nasal cavity. (Redrawn from K. P. Zhao et al., Effect of anatomy on human nasal air flow and odorant transport patterns: Implications for olfaction, *Chemical Senses* 29 [2004]: 365-379)

these airflows have increased. Some have focused on the normal nose, others have studied the effects of disorders such as polyps on olfactory function, and others have compared humans with dogs and other animals.

An example is shown in figure 3.2, from experiments by Pamela Dalton and her colleagues at the Monell Chemical Senses Center in Philadelphia (whose work was published as Zhao et al.), in which the system of turbinates inside the nose is seen in longitudinal display to show the pathways of the currents. The fastest flow is at the bottom of the nasal cavity, and the slowest is at the top in the narrow cleft just under the site of the smell receptors. The current flows at the top related to the smell receptors are composed of eddy currents due to the turbinates' complicated geometry.

Experiments with these models have shown several things of interest to smell function in general and wine tasting in particular. First, subjects show large differences in the shape and openness of their nasal cavity, giving rise to differences in current flows. Second, the two sides of the nose are not identical. This is a general finding among subjects. It could reflect the nasal cycle, the back and forth of congestion between the two sides that happens every 90 minutes or so, or real anatomical differences. Third, the flow lines are surprisingly rigid; air that enters at a specific site in the nostril is destined for a specific path through the nasal cavity.

For example, air flowing to the top of the cavity to stimulate the olfactory receptor cells enters at a precise site most anterior in the nostril. Finally, the authors show that the flow paths are highly sensitive to small obstructions and congestions.

This and similar studies have important implications for differences in smell sensitivity related to wine tasting, as indicated in this quote from Dalton and colleagues:

> Relatively small changes in the anatomy of the nasal cavity at specific locations can induce large changes in the airflow through and the odorant uptake on the olfactory mucosa. . . . Also, . . . a single cast or model to draw general gross flow features for the general population, would appear to be unsuitable for investigations related to human olfaction, since inter-individual variations in nasal anatomy and even anatomical variations of the same subject at different time frames due to congestion or decongestion can cause substantial variations in nasal air flow and odorant transport pattern. . . . At different anatomical conditions, air flow rate through the olfactory region and odorant uptake rate in olfactory mucosa can possibly differ up to two orders of magnitude. Finally, . . . anatomical variations can play a different role during inhalation and exhalation.

These results provide strong evidence that each of us varies inside our noses as much as we vary in our bodies—even, at different times, within the same body. These are called *interindividual* as well as *intraindividual* differences. To these differences in flow patterns within the nose are added the differences in our brains' responses. It is a wonder that we can reach any kind of agreement when assessing the aroma of wine!

CHAPTER FOUR

The Pathway for Retronasal Airflow

The pathways followed by retronasal air currents have not received as much attention as the orthonasal currents. We introduce them here to provide a basis for understanding retronasal smell and the internal wine aroma covered in chapters 13 to 15.

Retronasal Airflow Patterns in the Nose

With orthonasal smell, we saw that air sniffed in flows through the nasal cavity, generating eddy currents that carry the odor molecules up to the smell receptor cells at the top of the cavity. While many laboratories have documented these patterns, few have studied the patterns when breathing out. One such study, carried out many years ago in human cadavers by Arthur Proetz, showed that the two patterns are quite different. To summarize, orthonasal air passes in through the relatively narrow nostril, through the relatively large nasal cavity, and out through a relatively large opening (choana) at the back into the nasopharynx and upper airway. By contrast, retronasal air moves out from the nasopharynx through the large choana into the nasal cavity and gets backed up behind the nostril, causing turbulence that carries volatiles widely over the olfactory epithelium.

In their recent work on air currents in the nose, Pamela Dalton and colleagues at the Monell Chemical Senses Center reported:

A study of the streamlines . . . during exhalation shows a smooth air-flow pattern with no flow separation or re-circulation in the nasal airway. The regulatory effect of the nasal valve on airflow pattern and distribution in the nasal cavity is most pronounced during inhalation when it is at a location upstream to most of the flow. This finding sug-gests that reported differences between orthonasal and retronasal ol-factory sensitivity among some subjects . . . can be partly attributed to the different airflow pattern during inhalation and exhalation.

Current studies are underway by a team led by Sophie Trastour and Michael Brenner and their colleagues at Harvard University. In a brief report, they have obtained CT scans of human nasal cavities and have carried out computational analyses of the fluid dynamics simulations in the orthonasal and retronasal directions. From this, they have derived scaling laws for the rate of odor deposition under different flow para-meters. The results suggest that the internal geometry of the human na-sal cavity enhances retronasal smell. We have already noted that the proximity of the pharynx to the nasopharynx and nasal cavity appears to enhance retronasal smell in humans. The new evidence extends the idea that humans are adapted for the enhancement of retronasal smell. As we have noted, this presumably reflects the survival advantage of highly flavored and easily digested food in the human diet, as Richard Wrangham proposes in his pioneering book *Catching Fire: How Cooking Made Us Human*. It also adds to the evidence regarding the many adap-tations in the human head for consuming food and producing flavor, as Daniel Lieberman summarizes in his comprehensive book *The Evolu-tion of the Human Head*.

Experiments in laboratory animals have suggested that because of this direction of flow it may be possible for odor molecules to be ab-sorbed in a sequence across the olfactory receptor sheet, resembling the sequential deposition of components in a gas chromatograph, with this pattern as the basis for odor discrimination. This is the chromatographic theory of odor discrimination discussed in chapters 3 and 8. However, a reverse pattern would be produced during retronasal smell. This ques-tion deserves more experimental study of the eddy currents involved. At present, it seems more likely, as with orthonasal smell, that the differential activation of olfactory receptor cells is dominated more by the differing specificities of their receptor molecules to odor molecules than the patterns of absorbance in the epithelium.

In summary, it appears that the flow patterns in the nose are different for retronasal and orthonasal smell. This could produce differences in encoding the smell molecules, and differences in the perception of smell, but more research is needed.

Retronasal Airflow in the Pharynx

This brings us to the most difficult and least understood dynamic function in wine tasting: the contribution of retronasal smell.

To reiterate, breathing in is called orthonasal airflow because it is what most people consider the "normal" direction for filling our lungs and at the same time stimulating the sensory cells inside our nasal cavities to produce the sense of smell. Breathing out is called retronasal airflow because it is "retro," against the "normal" direction. But of course, as we have discussed, breathing out is just as normal as breathing in. The two always go together. We could as well call them "in-nasal" and "out-nasal" to indicate that both are equally normal, just in opposite directions.

Since smell is so important for most animal behavior, one might assume that other animals are better at it than we are, which would imply that retronasal smell might be less important for humans. However, in *Neurogastronomy*, I compared the pathways for olfaction in a dog and a human. The long snout of the dog, the highly convoluted nasal passageway, and the large extent of smell receptor cells are what one would expect of an animal with a highly sensitive sense of smell. But this is particularly relevant to sniffing in, which of course is what a dog is really good at when it is following tracks on the ground or meeting another dog. However, when it comes to eating, dog owners know that a carnivore such as a dog literally "wolfs down" its food, giving relatively little time to tasting the food and therefore sensing its internal smell. This goes with its relatively long and narrow nasopharynx, which is not optimally adapted for retronasal smell. This is not to say that dogs and other animals are not attracted to flavorful human food, as anyone with a dog that hangs around the dinner table well knows. However, in the wild, rapid consumption is usually a priority.

Fortunately, we are not exclusively carnivores; our ability as omnivores to eat almost anything makes wines and wine tasting possible. In contrast with other animals, humans have lost their snout. Our noses are small, the nasal cavity is short, and most importantly, the back of the mouth connects to the nasal cavity by a relatively short and wide nasopharynx. The volatiles released at the back of the mouth have relatively rapid access to the smell receptors in the nasal cavity, which also applies to the volatiles left coating the throat after swallowing. This means that conditions are enhanced for smell to make a major contribution to the taste of what is in our mouth and throat; in other words, conditions are enhanced for wine tasting.

The volatiles are carried in the air that is breathed out during exhalation. Their effectiveness depends on exhalation during respiration, as we have discussed. And yet, we are totally unaware of this contribution. We will see how this creates one of the greatest illusions in human behavior when we discuss how the sensory systems create the perception of flavor.

The Nose-Pinch Test: Internal Smell from Mouth to Nose

How can molecules in the mouth get to the nasal cavity to stimulate the smell receptor cells? The most immediate direct route is from the back of the mouth past the soft palate and up through the nasopharynx to the nasal cavity. The nose-pinch test reveals this route. In this test, which is often performed for schoolchildren to demonstrate the fun of doing science, a piece of candy is handed out, the nose is pinched, and the candy is placed on the tongue while breathing through the mouth. If the pinch is tight and the breathing calm, the subject reports sensing only sweetness on the tongue. When the pinch is released with the mouth still closed and breathing through the nose resumes, the subject reports a flood of flavor from the candy. Where does this come from?

Subjects are usually delighted, often remarking that they now understand why they do not taste much when their noses are stuffed up by a cold. However, the real lessons are as follows: First, most flavor comes from the aroma as sensed by smell. Second, this flavor comes only when breathing out through the nose. Third, if it occurs without the subject swallowing, it demonstrates that there is a route from the back of the mouth directly through the nasopharynx to the nasal cavity. Fourth, it is a route that acts quickly.

The response time to a drop of liquid chocolate on the tongue was measured by Yuri Masaoka and his colleagues at the Showa University School of Medicine in Tokyo. They used electrical waves recorded through electrodes on the head (the electroencephalogram [EEG]) to measure this response time following the onset of expiration after releasing the nose-clip, which was much more precise than the verbal report of the subjects. The results showed that the EEG response began to occur within 200 milliseconds (two-tenths of a second) of removing the nose clip. This may be taken as a measure of the time needed for expiratory air to entrain the odor molecules on the tongue and transport them to the nasal cavity, plus the time for the molecules to be absorbed in the mucus, stimulate the olfactory receptor cells, and begin to elicit the brain response.

The responsive regions in the brain were identified using brain imaging, beginning with the left entorhinal cortex and left amygdala at 200 milliseconds after expiration onset, the left entorhinal cortex and left hippocampus at 220 milliseconds, the left amygdala at 240 milliseconds, the left hippocampus and left medial orbitofrontal cortex at 260 milliseconds, and the left medial orbitofrontal cortex at 280 milliseconds. Interestingly, this was similar to the sequence observed with orthonasal stimulation in a previous study. The authors suggest that conscious perception begins with the response of the orbitofrontal cortex; as discussed in chapters 12 and 15, this is a region where taste input is integrated with smell to produce flavor. Since the receptor cell response by itself is believed to take up to 200 milliseconds, it indicates that retronasal smell from the mouth to the nose occurs extremely rapidly. This route is shown in figure 4.1.

These results show that there is a direct pathway for retronasal smell from the mouth to the nasal cavity. It may be noted that the studies cited have supported Paul Rozin's suggestion that in moving volatiles from the back of the mouth to the nasal cavity, retronasal smell occurs mainly by airflow rather than diffusion.

The Internal Route for Retronasal Smell

Although it is clear that retronasal smell makes a major contribution to the flavor of food and the taste of wine, the internal route the volatiles take is controversial. The traditional view is that because the wine is liquid, the back of the mouth needs to form a tight seal to keep the wine in while its taste and mouthfeel are sensed. In this view, retronasal smell therefore has to occur after swallowing, carried from the wine left coating the pharynx at the back of the mouth. This idea was proposed years ago by the studies of Burdach and Doty, the experts on deglutition at Rochester University (chapter 1).

The alternative idea, suggested by the nose-pinch test and the previous discussion, is that the seal can be lifted so that volatiles can pass directly from the back of the mouth through the nasopharynx to the nasal cavity during intermittent exhalation.

This question was addressed in 2002 by Andrea Buettner and her colleagues at the German Research Institute for Food Chemistry in Garching, Germany, using magnetic resonance imaging (MRI) and videofluoroscopy. They confirmed that the back of the mouth and the palate can form a seal that can be lifted:

> Based on these observations, a more intense flavor perception can only be achieved, when deliberately opening the velum-tongue border as done e.g., by winetasters. However, this deliberate opening is only possible either by placing a small portion of liquid into the oral cavity or by bending the head forward so that no liquid can flow into the pharynx.
>
> Note the reference to winetasters. It suggests that winetasters can train themselves to take small sips, bend forward, and consciously hold the back of the tongue away from the soft palate.
>
> Visualization of the oropharyngeal performances during consumption of liquid and solid foods revealed a direct link between dynamic retronasal aroma perception and intermittent openings of velum–tongue and velum–pharynx connections during the eating process. . . . we have to differentiate between minor retronasal aroma perceptions during the short-time openings of the velum–tongue border during mastication and the aroma pulses induced by each single swallowing event.

The more vigorous the chewing motions of the jaws, the more it stimulates associated openings of the palate:

> During mastication, an alternating series of open and closed stages of the velum–tongue border takes place . . . the extent of the opening being highly dependent on the intensity of the movements of jaw and tongue.

The authors further questioned how much of an odorant in the mouth is absorbed by the oral membranes (mucosa) compared with the amount spat out in the saliva. The results showed that the mucosa adsorbed (or resorbed) almost one-third of an odorant. This is available to produce volatiles in addition to the adsorption on the oropharynx after swallowing. These experiments also provided evidence that no enzymatic degradation of the odorants occurred after exposure to saliva.

Aeration of Wine in the Mouth

In wine tasting, we take advantage of this direct route when, with the wine in our mouths, we open our lips and suck in air. This is usually regarded as a move to "aerate" the wine, mixing in air to enhance the release of volatiles. The subsequent enhancement of the aroma of the wine follows an indirect route. While sucking in the air, the volatiles are picked up and carried through the back of the mouth into the oropharynx and into the trachea and lungs. No sensing of the aroma therefore occurs at this stage. Upon subsequent expiration, however, the volatiles are picked up from the aerated wine at the back of the mouth and transported to the nasal cavity by the retronasal route.

This requires a certain amount of practice because inhaling air leaves the glottis open, so care must be taken not to draw wine into the larynx and choke. At the same time, there is no stimulation of the nose until one breathes out to produce retronasal smell and the aroma of the wine. This maneuver of aeration is especially important during a wine tasting involving many wines, during which each wine is spat out to avoid the cumulative effect of swallowing too much. Aeration of the wine in the mouth enables the wine taster to substitute for the loss of sensing the aeration of the wine coating the throat after swallowing. Jean-Claude Berrouet demonstrated his expertise in this maneuver during our tasting, as mentioned in the appendix.

Enhanced Airflows from Mouth to Nose

What is the route for the volatiles in retronasal smell? It seems obvious that it is from the back of the mouth into the oropharynx and through the nasopharynx to the smell receptors in the nasal cavity, as indicated in figure 4.1, but no one had ever demonstrated it. We tested this route by quantitatively examining a fluid dynamics model of airflows through the throat in both the orthonasal and retronasal directions.

To do this, I sought a collaboration at Yale with an expert in fluid dynamics, Nick Ouellette, and a postdoctoral fellow with him, Rui Ni. This led to an exciting collaboration between four laboratories. We first obtained a fluoroscopic image of the throat area of a subject from clinical colleagues Mark Michalski and Elliott Brown at the medical school. Joseph Zinter and his student Ngoc Doan at the Center for Engineering Innovation and Design constructed a three-dimensional printout of the image. Nick and Rui then built a physical model and carried out tests of flow patterns in both the orthonasal and retronasal directions, and I worked with them to analyze the patterns.

The study showed a surprising result. During simulated expiration, the particles representing volatiles collect at the back of the mouth in a virtual cavity and are entrained by the flow outward toward the nose; by contrast, during simulated inspiration the flow goes inward with little entrainment of the volatiles (figure 4.1). The model thus suggests that even with normal slow breathing the specific anatomy of the throat is adapted for efficient retronasal smell directly from the back of the mouth.

We suggest that this route from the back of the mouth into the exhaling air plays a significant role in food flavor. A liquid, such as wine, poses the challenge of keeping it from leaking out of the back of the mouth while still letting it be picked up by expiring air. But according to recent studies, as we have seen, the soft palate can be trained to stay open until just before swallowing, allowing retronasal smell to act as long as possible.

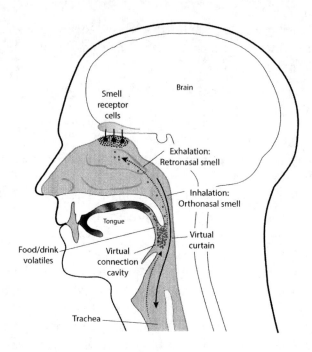

FIGURE 4.1

Diagram showing how food/drink volatiles appear, in a fluid mechanics model, to pool at the back of the mouth and may be differentially taken up by retronasal but not orthonasal airflow. (Based on R. Ni et al., Optimal directional volatile transport in retronasal olfaction, *Proceedings of the National Academy of Sciences USA* 112 [2015]: 14700–14704)

Tips for Wine Tastings

Wine books, of course, place a special emphasis on the methods learned from long experience for enhancing wine tasting. The points we have raised thus far come from a new direction: analyzing the fluid mechanics of the wine and the volatiles, with an emphasis on retronasal smell. It seems there is good agreement on the essentials, with new insights into the underlying mechanisms, as well as a basis for bringing the sensory experience into the picture.

Based on these aspects of flow dynamics and retronasal smell, the implications for the best strategy for serial tasting at a wine-tasting event include the following points. Start by taking moderate sips, keeping the sipped wine in the anterior part of the mouth to sense before and after mixing with saliva. Maximize the stimulation of mouthfeel and taste by vigorous agitation of the wine in the mouth. Sense the astringency and push the tongue against the hard palate to increase the release of volatiles. Maximize the aroma by keeping the back of the mouth open to facilitate retronasal smell. Let wine leak over into the valleculae in order to get a hint of an aroma burst from the throat. Spit out the wine to limit ingestion of the alcohol which dulls the analysis of the wine taste. Do frequent swallows of the remaining wine plus saliva for more prolonged retronasal testing of the finish from the wine coating the mouth and pharynx. (These steps can also enhance your enjoyment of wine at a meal, in which case you can skip the spitting out part!)

Perceiving the aroma both before and after swallowing is particularly important when tasting many wines, such as during the New York Spectator Wine Tasting that my wife, Grethe, and I attended at the invitation of Marilisa Allegrini of Amarone Valpolicella. The rule at such a wine tasting, with wines presented for tasting at more than 200 booths, is to take a small sip, evaluate the taste, spit it out, and move on to the next. Enhancing as much as possible the direct route by the means outlined previously enables the taster to evaluate (and enjoy) the wine even with limited benefit of swallowing it.

There are thus three internal routes for volatiles to pass from the mouth to the nose to add aroma to the flavor of wine. With the wine in the mouth, a direct route leads from the back of the mouth to the nasal cavity. An indirect route is aerating the wine with subsequent expiration. Finally, there is the all-important indirect route from the mouth to the throat to the nose. This occurs after swallowing, an action we usually take for granted. However, swallowing is critical to wine tasting and therefore deserves a chapter of its own.

CHAPTER FIVE

Swallow, Aroma Burst, and Finish

Swallowing is one of those automatic housekeeping functions we do without thinking about it, to clear the saliva we produce throughout the day, as well as food and drink. As such, it seems a rather humdrum activity; how could it be important for a sophisticated function such as evaluating and enjoying a wine? Ironically, however, swallowing is one of the most crucial functions for wine tasting.

Surprisingly, it was one of the first bodily functions studied by early physiologists. In 1836, François Magendie in France reported that swallowing consisted of three parts: an initial stage of preparing the food in the mouth; a second stage of moving the food into the pharynx, and a third stage of moving the food into the esophagus. Around 1900, it was recognized that the driving force came from a central pattern generator in the brain. In the mid-twentieth century, Robert Doty and his colleagues studied the sequence in detail.

From the 1990s on, swallowing has been studied in detail, and the neural circuit has been identified and its function characterized for each stage. Most of the studies have focused on swallowing foods, in which an initial ingestion of food is chewed into a soft bolus and eventually into a more fluid mash before swallowing. We will interpret the implications of those studies for swallowing a liquid, such as wine.

The initial steps within the mouth are under voluntary control, up to when we decide to swallow. Swallowing is a highly complex and integrated act. When drinking wine, most people consider swallowing as an end to the wine experience. But to an expert, swallowing sets the stage for a crucial part of wine tasting: delivering the wine into the pharynx, from which it passes to the esophagus but leaves behind a coating of wine that is a reservoir for the strongest stimulation of retronasal smell.

Since swallowing occurs largely automatically, we give it scarcely a thought, but in fact, like so many things controlled by our nervous system, it is a complex act. There are investigators who devote their lives to studying the physiology of swallowing. Why should this be, and is it of practical importance? A moment's reflection indicates how important it is. Disturbances of the swallowing mechanism are involved in choking, which is the fifth most common cause of accidental death in the United States, accounting for some 2,500 deaths annually. Most of these are in young children, reflecting the time it takes for the complex mechanism to mature. Because it is so complex, it is a behavior that varies among different people. Individual variation is the basis for natural selection in evolution—and also in abilities in wine tasting.

The Swallowing Muscles

Swallowing involves some 25 muscles of the jaw and neck. Several of the key muscles attach to the hyoid bone (*hyoid* is Greek for "shaped like the letter *U*") (figure 5.1).

The hyoid is a slender bone at the base of the mouth (derived from the first gill arch of our long-distant fishlike antecedents), just above the Adam's apple. The set of muscles shown is situated above the hyoid, to control what is going on in the mouth. These so-called *suprahyoid* muscles consist of the *digastric* (which has two heads that pull it forward or backward), *stylohyoid* (attached under the temporal bone, it pulls the hyoid back), *mylohyoid* (arising from the line of molar teeth of the mandible), and *geniohyoid* (from the Greek *genio* [chin], so this runs from the back of the chin to the hyoid). This is more than most readers need to know, but there it is for the experts.

To take the first step in swallowing, the anterior parts of these muscles contract to raise the hyoid bone and the base of the tongue. This squeezes the wine toward the pharynx. Then comes the clever part. In the second phase, the posterior parts of the muscles contract, lifting the hyoid bone up and backward. This pulls the thyroid cartilage of the larynx

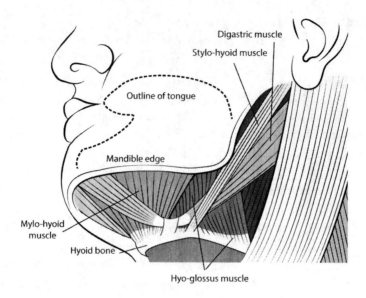

FIGURE 5.1

Some of the key muscles that control the hyoid bone and make swallowing possible. (Adapted from H. Gray, *Anatomy of the Human Body*, 26th ed., ed. C. M. Goss [Philadelphia: Lea & Febiger, 1954], 437, fig. 441)

upward, which causes the epiglottis to flop down upon the larynx, closing off the trachea. The wine is now clear to move out of the back of the mouth through the oropharynx and into the esophagus (figure 5.2). Comparative anatomy teaches us that this intricate mechanism has emerged over millions of years of evolution from the gill arches of fish to us.

The Swallowing Central Pattern Generator

The circuit that controls this swallowing action is closely associated with the cluster of cells known as the *nucleus of the solitary tract* (NST), which is the main relay station for sensory signals from the taste buds on the tongue, as we shall see in chapter 12. Presumably, this sensory information is the most relevant to monitoring not only the wine's taste but also the wine's location over the tongue.

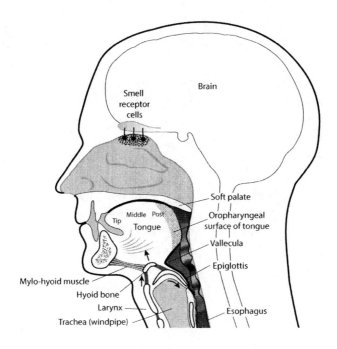

FIGURE 5.2

During swallowing, raising the hyoid bone, causes the epiglottis to flop down and close off the larynx, while the soft palate closes off the nasal cavity, so that the wine can flow through the throat and into the esophagus. This leaves a coating of wine on the throat that contributes to the aroma burst of retronasal smell.

Swallowing is such a critical function, with the risk of choking, that the swallowing circuit acts in a very programmed way. We have seen that many such vital functions are found in the body (chewing, respiration, heart rate, locomotion, etc.), and that the circuits that control them, *central pattern generators* (CPGs), are situated in the brainstem where they can act more or less automatically. The circuit controlling swallowing is therefore called the *swallowing central pattern generator* (sCPG).

Like other central pattern generators, the sCPG has three parts: the sensory inputs that trigger it, the local interactions that generate the rhythm of muscular control, and the outputs to the muscles. In simple terms, there are two groupings of neurons. The *dorsal swallowing group* (DSG) includes part of the nucleus of the solitary tract, the relay center for the taste pathway. It is activated by inputs from the taste receptors and also by conscious commands from the cerebral cortex. Its neurons generate a pattern of activity programmed for the sequential activation of the swallowing muscles. They connect to a *ventral swallowing group* (VSG) that receives the dorsal output and distributes it to the muscles.

Regarded as a functional entity, the sCPG includes neurons in parts of the brainstem that contribute not only to swallowing but also to functions such as movements of these and related muscles in forming speech, in breathing, and in vomiting. These neurons are flexible in these other behaviors as well. For neuroenology, the take-home message is that swallowing is due to a neural circuit that, even though it is far from the highest centers, is one of our most complex systems. We will have evidence of this in describing the orchestration of swallowing wine.

The Orchestration of Swallowing

You can recognize from your own experiences that there are roughly three main stages of swallowing (you can follow this sequence by referring to figures 3.1 and 5.2).

Oral Stage of Swallowing

Swallowing begins with the brain, with the prefrontal cortex at the center of the command circuits making the decision to initiate the process. These inputs to the brainstem, including the sCPG, begin moving the liquid toward the back of the mouth through the initial forward and upward action of the suprahyoid muscles. The uvula (soft palate, or velum) moves upward to begin closing off the nasopharynx, and the epiglottis begins moving downward to close off the trachea and prevent choking.

Pharyngeal Stage of Swallowing

Studies show that a bolus of food is propelled from the back of the mouth into the pharynx by the backward contraction of the suprahyoid muscles and the upward contraction of the back of the tongue to push the food away, aided by relaxation of the pharynx to receive the bolus. Similar actions appear to occur with liquids so that wine passes from the back of the mouth into the pharynx. If you hold your throat lightly just above the Adam's apple, you can feel it lift at the moment you swallow. The combined actions of the soft palate above and the epiglottis below seal off respiration. This momentarily interrupts airflow and breathing, so there is no expiration and no stimulation of retronasal smell by any means.

Esophageal Stage of Swallowing

The final stage occurs when the wine passes into the esophagus. As soon as this happens, the soft palate relaxes and the epiglottis lifts so that respiration can resume. The motor neurons to the muscles of the esophagus are activated to initiate the wave of muscle contractions (*peristalsis*) that move the wine through the esophagus, through the esophageal sphincter, and into the stomach.

Postswallowing

The Aroma Burst

After swallowing, some wine is left coating the mouth and throat. The coating left by foods after swallowing has been the subject of research (described in chapter 7), which has characterized the coating in terms of components that constitute a *matrix* and components giving rise to the *volatiles*. The matrix keeps the coating sticking to the throat walls, which allows the volatiles to be released into the respiratory streams as soon as respiration resumes.

Although it has received little attention, this mixture of matrix and volatiles, together with the saliva from the mouth, is crucial for the wine-tasting experience (see box 7.1). The concentration of volatiles is of course at its peak with the first expiration after swallowing. This gives a strong sensation of the wine aroma accompanied by a strong perception of the wine "taste." Some authors have called this initial strong perception the *aroma burst*. For many tasters, it is the strongest contributor to the taste of wine.

The Finish

This mix of matrix and volatiles varies with different wines. It becomes increasingly complex with aging of the wine, as Jean-Claude Berrouet demonstrated to me in tasting a twelve-year span of Petrus wines (see the appendix). Over a period that can last many minutes, the coating on the mouth and throat releases its volatiles into the adjacent stream of air. During exhalation, this air is transported to the nasal cavity to stimulate retronasal smell. During inhalation, the air must be carried into the trachea and on into the lung; little is known about its fate.

After the aroma burst, the lingering stimulation the volatiles provide with continued cycles of exhalation is called the *finish* of the wine in English; in French, the *longueur*. We will see that the richness and the duration of this period is a crucial component of how a wine is characterized and rated. (Of course, many still think it is due to taste, even though it is coming from retronasal smell.) With repeated swallowing, volatiles accumulate along the entire airway and extend the contribution of retronasal smell to the prolonged finish.

BOX 5.1
Factors Affecting the Finish

Vapor pressure differences between aromatics
Relative absorption in and/or destruction by saliva enzymes
Rates of absorption in the pharynx and nasal mucosa
Destruction by mucosal enzymes
The reversible association of aromatics in nonvolatile complexes with
    wine constituents
Learned associations between taste and aromatic compounds

R. S. Jackson, *Wine Tasting: A Professional Handbook*, 2nd ed. (Amsterdam: Elsevier, 2009), 96.

Ronald Jackson, author of *Wine Tasting: A Professional Handbook*, states that the most significant contribution of retronasal olfaction in wine tasting is its involvement in the perception of the finish. According to Jackson, the perceived duration and intensity of the finish depend on multiple factors, summarized in box 5.1. These factors are discussed at various places in this book.

———————————

To summarize, studies of subjects drinking fluids show that with the fluid held in the mouth, retronasal smell can occur through the back of the mouth and the open velum. After swallowing, the coating of the wine on the throat has a direct input to exhaled air through the unobstructed nasopharynx. Repeated swallowing thus contributes to the sustained activation of the olfactory receptor cells through retronasal smell, providing the most powerful input for sensory discrimination of the wine.

PART II

How Sensory Systems Create
the Taste of Wine

One might think that an account of the sensory systems creating the taste of wine would start with the taste system. Ironically, taste as a sensory system actually plays a limited role in wine taste. By contrast, smell plays a large role through both the orthonasal and especially the retronasal routes. But before either taste or retronasal smell can begin, the first sensation we have from the wine in our mouths is the touch of it against our lips, mouth, and tongue. But even before that, we look at the bottle and examine the color of the wine, which makes a deep impression on our expectation of its taste. There is a lot going on, which is why it can be claimed that the flavor of wine, like the flavor of food, engages more of our brain than any other human behavior.

In keeping with the functional approach we have used to the biomechanical steps in wine tasting, we will follow the sequence of activation of these sensory systems that occurs in tasting a wine.

1. Vision comes first; we see the wine in the bottle and then in the glass and examine its color.
2. We test its bouquet in the wine glass by breathing in to activate our smell receptors by orthonasal smell.
3. The wine goes into our mouths, activating many kinds of touch receptors and making everything that follows seem as if it is coming from the mouth.
4. The wine activates the taste buds on our tongues and upper throats to produce the specific sensation of taste.

5. The all-important activation of our smell receptors by retronasal smell adds the richness of the aroma to the wine taste.
6. The sounds of the swishing of the wine in our mouths and the sounds of swallowing are part of the wine taste.
7. After swallowing, we sense an aroma burst from the volatiles left coating the throat, with the first expirations providing the strongest stimulation of retronasal smell.
8. Finally, we sense by continuing retronasal smell the lingering finish coming from the volatiles on the surfaces of our mouths and throats.

We discuss each of these steps in the ensuing chapters and see that wine is a true multisensory experience, stimulated and coordinated by the fluid dynamics of the wine and volatiles covered in part I.

CHAPTER SIX

Sight
Creating the Color of Wine

Except for those who are blind and during blindfolded tastings, the experience of wine begins by seeing the wine before we drink it. With food, it is said that "we eat with our eyes"; with wine, we can say that "we taste with our eyes." As with every food and beverage, the product is marketed to have a maximum appeal to the consumer. This involves all the strategies of advertising, such as a convenient package, attractive labeling in bright warm colors like red or yellow, a good feel when holding the product, and a nice smell upon opening it. Eons of advertising have shown that all of these factors can have a strong effect on our evaluation of and desire for the product.

The same is true of wine. The common bottle size containing 750 milliliters (about 1½ pints) is convenient to market and serve—a half bottle is too small, a double size (*magnum*) is too unwieldy, for most groups. The wine label is important, to project a high quality to the consumer, with an appropriately traditional label signifying origins going back centuries or a modern label for a new wine. The various shapes of bottles may be a factor in bonding with wines from different regions and of different types.

The most important visual characteristic of wines is color. As everyone knows, the world of wines is mainly divided into whites and reds, along with a fainter red called *rosé*. Whites are not actually white: "pale" would be a better term.

Whites are fermented mostly without the grape skins. In taste, they tend to lack the strong flavors that the molecules of the skins usually produce in reds. Whites are usually served cooled or chilled, which further limits the effects of the taste molecules on the taste bud receptors and the volatiles released into the orthonasal and even retronasal air. Their color is pale because the wine is composed mostly of the flesh of the grapes. By contrast, red wines are fermented with the grape skins, which contain anthocyanins and other molecules that reflect light in the longer wavelengths.

Creating Color

We have said that taste is not in the wine; it is created by the brain, and the same goes for color: "red" and "white" color are not in the wine but are created by the brain from different wavelengths of light. Color is important not only for distinguishing between red and white (and in between, rosé) wine but also for seeing fine distinctions between different wine varietals and different ages. In addition, we will see that color makes a critical contribution to our initial observation of the state of the wine in the glass.

It is worth noting that from an evolutionary perspective, humans are lucky to be able to distinguish colors. Most of our mammalian cousins have only one kind of photoreceptor, meaning that they can only see dark and light; your pet cat, for example, has very limited color vision, and your pet dog has none. Our primate cousins and human ancestors were fruit eaters; when during evolution millions of years ago they reconstructed color vision, it enabled them to distinguish the colors that indicate ripeness in plants and fruits. For modern humans, wine making and wine tasting without color vision are unimaginable. This is not so much about telling the difference between red and white wines as about distinguishing clues from the color about the state of the wine.

The mechanisms of vision make up a vast field. We will focus mostly on how vision works in creating our perception of red wine color.

There are three steps. The first is transforming the light stimulus entering our eyes into a photoreceptor response. The second occurs when visual circuits collect the receptor responses and construct a representation of the visual image. And the third is processing the image to create the conscious perception of the stimulus. These steps underlying perception in the visual system are built on principles that apply to all the sensory systems.

Transforming Light into Receptor Responses

When we see an object like a glass of wine, it is due to light of different wavelengths bouncing off it and into our eyes. These waves have no color; they are just constantly coming in at their particular wavelengths and frequencies, like radio waves and other signals.

The light waves are composed of charges called *photons*. The photons, focused by the lens of the eye, hit the membranes of two main types of receptor cells at the back of the retina. *Cones* (so-called because they have a conical shape) respond to different wavelengths in bright light. Cones are gathered tightly at the center of the retina to form a *fovea*, where they serve high-definition vision in bright light. There are three types of cone cells, each containing a receptor molecule called a *rhodopsin* that is tuned to respond best to light at short (S), medium (M), or long (L) wavelengths. When a photon hits a rhodopsin, a tiny molecule (*retinal*) inside flips into a different form, which sets up an electrical response in the cell called the *photoreceptor response*. By contrast, *rods* are scattered in the periphery of the retina and respond best to weak light at intermediate wavelengths. That is why we do not see color in the dark—and why it is necessary to observe the color of a wine under a bright white light.

Creating Color from Receptor Responses

For the second step, thousands of photoreceptor cells connect to circuits in the retina that encode the shape of the image as well as the different wavelengths coming from the image. The L wavelengths give rise to the perception of red, the M to green, and the S to blue (the peak for L actually gives rise to green-yellow, but it extends into the red) (figure 6.1). Note again that the light itself has no color, only different wavelengths; the brain creates the colors we perceive by its processing circuits. To simplify, we will use green and red to refer to the M and L photoreceptors.

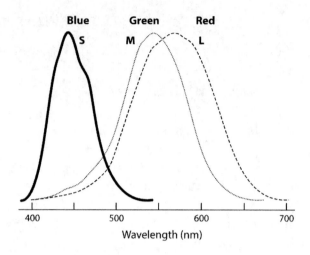

FIGURE 6.1
Circuits in the brain create the perception we call color from the different wavelengths of light: *S*, short; *M*, medium; *L*, long.

We have emphasized the color created by the peak sensitivity, but the figure shows that the receptor sensitivities extend beyond to overlap with each other, forming unique combinations of sensitivities for any given wavelength.

Of special interest to the wine taster is the region of the peaks of the *M* and *L* wavelengths, which give rise to green and red colors and the orange and yellow colors we perceive near the peaks. This is exactly the region where we make the most important distinctions about the slight differences in color that characterize the different varietal wines and wine ages. The peak sensitivities of the green and red cones are very close together (reflecting the fact that the green cone receptor genes were duplicated from the red cone receptor genes during the early evolution of primates several million years ago). The enhanced discrimination of wavelengths increases our ability to discern fine detail in our visual world, as you can see if you compare a color image with the same one in black and white.

The receptors are proteins composed of a chain of more than 300 molecules called *amino acids*. Research has shown that the green and red receptor molecules differ by only three amino acids at the critical sites in their protein chains. Think about how important this amazingly small difference is as you judge the subtle shades of red in the edge of a young wine as you tilt the glass or the faint shades of yellow or brown as a wine ages. There is considerable interest in developing quantitative measures relating the shift in hue to the shift from red anthocyanins to other molecules. The colors created by the brain thus track the "chemical age" of the wine, as first explained by Chris Somers and Michael Evans in the *Journal of the Science of Food and Agriculture* in 1977.

So now you know why the wavelengths of light bouncing off the wine you are examining need to be as pure as possible to indicate accurately the molecules responsible for the color you perceive.

Many factors can compromise that accuracy.

Of first concern is the ambient light. In a testing lab, the light is always pure white, avoiding ultraviolet or infrared wavelengths that would skew your perception toward red or green/blue. In your own testing situation, however, that might not be the case if it is a dinner with the lights turned low or a hall lit by fluorescent lights or the new energy-efficient type of bulb.

Second, you hold up the glass and tilt it to see through the edge of the wine where the color is most visible. In a testing lab, you see this against a white wall or a white table surface. At home or in a restaurant, a colored tablecloth or patterns in the wallpaper could compromise your testing of the wine color.

Third, the wine glass is also a factor. Professional testing is done with glasses suited to either a red or white wine so that when the glass is tilted the edge spreads out in an optimal manner. The glass, of course, needs to be spotlessly clean, without contributing to the color or the ability to see it. An old tradition in wine tasting is to swirl the wine in the glass and observe how it creates drip lines—"*legs*"—that are supposed to reflect the chemical composition of the wine. The more skeptical view is that this mostly indicates how clean the glass is.

Processing Circuits for Discriminating Color

The photoreceptors in the retina respond to the form and color of the wine in the glass, but these responses by themselves do not produce "color." Only the actions of nerve circuits create color. This starts with

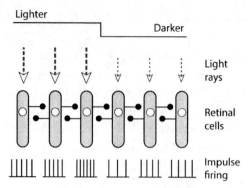

Lighter

Darker

Light rays

Retinal cells

Impulse firing

FIGURE 6.2

Lateral inhibition in the retina. At the border between differences in light stimulation (*downward arrows*), the inhibitory interactions between retinal cells suppress the responses of less activated cells (on the right), causing enhanced responses in more strongly activated cells (on the left).

the circuits in the retina. The response of the photoreceptors is transferred by *bipolar* cells to the *ganglion cells*, which carry the output of the retina to the brain. But within the retina are two stages of interactions. The bipolar cells interact with each other through *horizontal cell* interneurons near the photoreceptors and *amacrine cell* interneurons near the ganglion cells.

These connections do two things to the images. They enable more active ganglion cells to inhibit less active ganglion cells nearby; the mechanism is shown in figure 6.2. This is called *lateral inhibition*, which produces *contrast enhancement*; that is, it enhances the more active cells and depresses the less active cells. This enhancement of differences is a fundamental property of nerve circuits (and many other processes in the body). Increasing the contrast of the image is a more efficient way to process, send, store, and perceive the information. We will see this operation used in other sensory systems as well.

Since the peaks for green and red are so close together, how do we make the distinctions between them that are so important for wines (and many other objects)? The answer lies in the nervous circuits. The horizontal connections in the retina provide not only for enhancing contrast in space but also for enhancing contrast between wavelengths. In recordings from individual cells, it was found that a cell that responds

64

to a point of red light could be inhibited by a surrounding field of green. Similarly, other cells could be activated by a point of green light and inhibited by a surround of red. Thus was discovered the *color opponent* system, which enhances the ability to discriminate very small contrasts between green and red wavelengths in the region of red-orange-yellow-green (see figure 6.1). This is the neural basis for our ability to discriminate these colors when we examine the wine in our glass.

Conscious Perception of Wine Color

The output of the circuits in the retina is sent for higher processing to a relay station called the *lateral geniculate nucleus,* a part of the *thalamus,* which is the gateway to the neocortex. The neocortex is the "new cortex," the great rumpled sheet of nerve cells that covers the brain. Created by the earliest mammals over 200 million years ago, it contains the neural circuits that are critical for the higher brain functions that reach their peak in humans.

The visual image arriving there is not like a photographic image that retains every detail of the scene. At every stage of processing, the image is extracted to heighten the aspects of the object that are most significant for the behavior of the beholder. What we care most about in our visual world are things that are changing, so our nerve cells do not even register the static parts of the image, like the room around us or people standing still nearby. However, items that are changing, like someone moving or a tilting or swirling wine glass, are heightened.

Out of the differing responses of the photoreceptors and the processing circuits at the cortical level, the *perception* of color is created. This is why we say that color is not in the object; it is created by the brain. The earliest stage at which conscious perception of objects can occur is in the sensory qualities, such as color, that are called *qualia.* One of the great challenges in neuroscience and psychology (and philosophy) is to understand how our brains create qualia to represent physical properties.

The area that receives the fibers coming from the lateral geniculate nucleus is called the *primary visual receiving area* (V1). There are connections from here to higher-order processing areas (V2, V3, and above). These processing steps constitute the brain's way of representing wavelength and other visual attributes, such as movement, just as we will see that smell is the brain's way of representing odor molecules. Peter Gouras, an expert in vision, has said that qualia such as color are actually illusions because they are constructed by the brain from uncolored light waves.

We have seen that much of the importance of color to a taster of red wine is in the medium- and long-wavelength light that produces the shadings of red. The circuits producing these shades, by opposing responses between green and red wavelengths, start in the retina and maintain that distinction into the cerebral cortex. We use those circuits when we are discriminating between the reds of different wine varietals and ages. The red of a burgundy tends to be lighter than the deeper red of a Bordeaux. The red of a younger vintage tends to be lighter than the deeper, more complex red of a wine of advancing age.

Since the peaks of green and red responses are close together, discriminating between them requires several conditions. The first is the ability of the opposing circuits to create the perceptions of the colors and discriminate between them. The second is the power of our top-down circuits for intensely focusing our attention on those circuits. And the third consists of our memory circuits for matching the color perception to our experience of the range of varietals and vintages we remember. The brain uses similar circuits to make distinctions between fine details in our visual space, between closely spaced touch signals on the skin, or between odor molecules with similar odor-stimulating properties.

How Color Can Fool a Wine Taster

I learned about these differences between vintages of reds from Jean-Claude Berrouet in his Petrus testing laboratory. Most of us can see some of these differences, and experts have the ability to identify individual vintages and chateaus. However, a classic test in 2001 revealed that wine tasters can fail to discriminate between them. The test is described in detail in *Neurogastronomy*, so I will summarize it briefly here.

The experiment was carried out by Gil Morrot, Frédéric Brochet, and Denis Dubourdieu of the Faculty of Oenology at the University of

Bordeaux, deep in the heart of the Bordeaux wine region. The aim was to test subjects, using a white and a red wine, to see to what extent the wine's color influenced their evaluation of the odor. It seems obvious that, given the differences we have noted between the composition of red and white wines, differentiating between the odors would have been a simple task. However, on the basis of their long experience with wine tasting, the authors hypothesized that the color would dominate the odor in describing the wines. The red was produced from cabernet-sauvignon and merlot grapes; the white from sémillon and sauvignon grapes.

The test subjects were undergraduates in enology at the university, before their formal education began. Since we have already seen that noses vary greatly and will see that brains do too, the experimenters tried to reduce the individual variation by providing a limited set of descriptors that everyone could use to describe the wines—or they could use their own. The subjects then tasted the wines and used the descriptors to describe them and differentiate between them. A week later, they returned to repeat the tests, presumably to test for consistency in naming. A requirement, again to reduce variability, was to use the same set of descriptors they had used the first time.

Hidden from the subjects was the fact that the red wines the second time were all the same white wines, colored red. The comparisons were therefore being made between two sets of the same white wine but with one colored as if it were the original red wine. Nonetheless, the subjects evaluated the false reds using the same descriptors they had used to characterize the real reds the first time around.

How could this confusion have arisen? Morot and his colleagues offered several explanations. First, the subjects focused on using the descriptors they had used before, not using new descriptors which might have encouraged more diversity in their evaluations. Second, the authors suggest that the experiment tested the limits of language as much as the limits of sensory discrimination. Wines are described using language-based descriptors, which have complex cognitive contexts that condition sensory discrimination, as we shall see in chapter 18 on language in wine tasting. And third, the authors note that the way the brain is organized, with the orbitofrontal cortex, which mediates the highest cognitive functions, connected to the hypothalamus and other regions involved in emotional feelings about food, means that the two centers—cognitive and emotional—in a sense compete for every judgment they share.

In conclusion, judging the color of a wine takes place within a framework of integrating the senses, the emotions, the memories, and the use of language. It is a theme fundamental to the broad concept of wine we are calling neuroenology; a theme we will see repeated often as we assess the full range of brain mechanisms in wine tasting.

CHAPTER SEVEN

Orthonasal Smell
Wine Molecules Meet Smell Receptors

After examining a wine for its color and consistency, the next step in wine tasting is to lift the glass to the nose and sniff. This carries the volatile molecules released by the wine into our nasal cavity, where they stimulate receptor molecules in our olfactory sensory cells, which eventually leads to our perception of the smell.

The importance of smell to the flavor of wine cannot be overemphasized. In the authoritative *The Oxford Companion to Wine*, Jancis Robinson states, "The 'smell' of a wine may be its greatest sensory characteristic, but is also the most difficult of its attributes to measure and describe." The difficulty begins with the confusion in the terms used to describe smell. Robinson goes on to note: "In this book . . . the word flavour is used interchangeably with aroma." However, "flavor," in a broader perspective and in this book, refers to the combining of all the senses related to food and drink. And "aroma" is often inadequate to cover not only the multisensory perception of flavor but the myriad changes that take place in the release of volatile molecules from wine at the different stages of fermentation, maturation, and wine tasting.

Different terms have therefore evolved to apply to the perception of wine smell, which can be summarized as follows:

- "Smell" and "odor" are general terms for the overall perception of volatiles, whether pleasant or unpleasant.

- "Aroma" refers to the pleasant perception of volatiles as sensed either orthonasally or retronasally. Some use aroma to apply specifically to the smell emanating from a young wine.
- "Bouquet" can refer to the aroma of the wine in the glass. Some use it to refer to the aroma that occurs with aging.
- "Internal aroma" (our new term) will refer sometimes to retronasal smell from the wine in the mouth and throat.

For convenience, we will use "smell" as a general term, and "aroma" for perception of volatiles, unless referring to specifics of of a given stage in perception.

In this chapter, we explain current research on orthonasal smell, suggesting how, beginning with the odor molecules, the brain creates the perception of wine aroma while initially sensing the wine in the glass. We will cover the contribution of the internal wine aroma to the wine taste in chapters 13 to 15 on retronasal smell.

To explain the perception of smell, many books begin with the psychology of the wine taster. However, the science and art of neuroenology start with the volatile molecules in the wine (figure 7.1 and box 7.1).

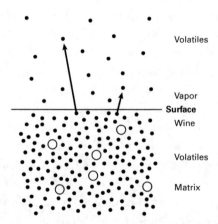

Volatiles

Vapor
Surface
Wine

Volatiles

Matrix

FIGURE 7.1

Schematic view of how volatile molecules are released from wine due to their vapor pressure. These are the volatiles sensed in orthonasal and retronasal smell. The length of the arrows indicates the vapor pressure of a given volatile molecule. Open circles are nonvolatile molecules that form the nonvolatile wine matrix.

BOX 7.1
Where Does Volatility Come From?

What makes a molecule in a wine "volatile" so that the olfactory system can sense it? Figure 7.1 shows a simple diagram of a fluid that contains many molecules: some are large and remain dissolved in the liquid (*matrix*). Others can be released into the overlying air if they are small in size and have enough energy to bounce out of the liquid—in other words, vaporize. These molecules are termed *volatile* (from Latin for "fly"). Each type of molecule has its own tendency to evaporate. The warmer the temperature, the more energy a molecule has to bounce out of the liquid.

It may be noted that the most common volatile acid is acetic acid, produced by yeast in the presence of oxygen. Very small amounts of acetic acid give wine an edge or "lift"; however, too much acidity results in vinegar. The volatility of a wine may thus have a negative meaning when this occurs.

Only the volatile molecules are involved in the steps of neural processing that create our mental images of the wine aroma. Volatile molecules vaporize from the surface of the wine in the glass, from the wine on the side of the glass after swirling, from the mouth, and from the pharynx after swallowing.

Temperature, which agitates the molecules and causes them to virtually bounce out of the liquid phase and into the air, increases vaporization. The tendency of a molecule to escape from liquid into air is called its *vapor pressure* (see figure 7.1). Different types of molecules have different vapor pressures. The contribution of each type of molecule to the mix is called its *partial pressure*. Initial orthonasal testing attempts to identify both strong and weak volatiles.

Sniffing the wine in the glass is both a science and an art, which is well described in most books on wine tasting—for example, Ronald Jackson's *Wine Tasting: A Professional Handbook*. At home and in restaurants, all too often the wine is poured to nearly fill the glass; as mentioned, this should be avoided, as it limits the headspace in the glass and consequently the bouquet, as well as eliminates swirling to increase the bouquet. A good glass will have a bowl shape to enhance the bouquet. Make use of it by pouring the wine to about one-quarter full to leave a large headspace for the aromatics to fill. First sniff without swirling, to sense the most volatile components, aiming the nose at the edge of the glass and then the middle. Wait to let the surface aromatics equilibrate and vaporize into the headspace again. Then swirl the wine to bring out the weaker volatile components. Note the different perceptions between quick sniffs and prolonged inhaling. Give time between each sniff for the olfactory sensory neurons to recover from adaptation, which lowers the sensitivity; 30 seconds of rest between sniffs is reasonable. For those interested, the effects of the shape of the glass on the surface area of the wine within and on the sides after swirling in relation to the volume of headspace gas and the diameter of the mouth of the glass have been calculated with mathematical precision. They are presented and discussed in Jackson's book.

Identifying Molecules in Wine Aromas

As we noted in chapter 2, the molecules that give rise to wine aromas are a small class among the thousands of volatile molecules that organic chemists have synthesized and studied. This has stimulated the development of increasingly sophisticated instruments that are important for the wine industry because of their ability to identify compounds responsible for wine aromas. One of these instruments, or rather set of instruments, is liquid chromatography together with mass spectrometry (LC-MS). Liquid chromatography involves a long tube through which an electric current is applied to move components in the fluid according to their molecular weights. Mass spectrometry then measures the amount of each component.

The components display as a series of "spikes." The placement of the spikes shows the different mass of the compounds, which can be identified by consulting libraries of compounds. The height of the spikes indicates the amount of each compound. By itself, this can be misleading because amount and aroma do not necessarily correlate: a substance may be present in small amounts but have a strong aroma or be present in large amounts and have no aroma at all.

Psychologists then test each spike for its odor, characterizing its strength and also its smell. This approach is called *LCMS-O*, developed by Terry Acree and his colleagues at Cornell University in 1984. LCMS-O is a powerful technique for beginning to identify the contributions of different types of molecules to the complex aromas of wines. Eventually, it may be possible to make profiles for each varietal and even for individual estates. However, because of so many variables, this goal is still a long way off.

This gives only the barest view of the increasing role that chemistry and psychology (and by implication, brain physiology) together are playing in prying apart the components of wine. This is stated clearly in a recent review (box 7.2).

BOX 7.2
Chemistry and Wine Flavor

Over the past century, advances in analytical chemistry have played a significant role in understanding wine chemistry and flavor. Whereas the focus in the 19th and early 20th centuries was on determining major components (ethanol, organic acids, sugars) and detecting fraud, more recently the emphasis has been on quantifying trace compounds including those that may be related to varietal flavors. In addition, over the past 15 years, applications of combined analytical and sensory techniques (e.g., gas chromatography-olfactometry) have improved the ability to relate chemical composition to sensory properties, whether identifying impact compounds or elucidating matrix effects. Many challenges remain, however . . . some of the recent research [is] aimed at understanding how viticultural and enological practices influence grape and wine volatiles . . . linking composition to sensory properties . . . [and] linking grape, yeast, and human genomics to wine chemistry and flavor.

From Eberler and Thorngate, 2009.

Odor Molecules Are Among Nature's Most Complicated Stimuli

What kind of stimuli are these many different kinds of molecules? Here, we need to take a step back and compare these molecules with other kinds of stimuli to get a better idea of why they are so complicated.

Analysis of most sensory systems starts with a clear understanding of the stimulus. Most senses are *unidimensional*; in other words, they vary along only one dimension or are otherwise specific in nature. Vision has light waves that are shorter or longer. Hearing has sound frequencies that are lower or higher. Somatosensation has touch that is localized to specific sites in space. Taste has its specific modalities. Along with this simplicity, each stimulus is identifiable and can be controlled so that the experimenter knows exactly what is being delivered.

Smell is different. All the smell molecules illustrated in box 7.3 are small, so they can vaporize into the air. But they differ in many ways, beyond their molecular weight, so they are what we call a *multidimensional* stimulus. They do not vary along just one dimension, such as number of carbon atoms. They vary along many dimensions, such as different functional groups, sites of double bonds, sites and types of side groups, aromatic side groups, and multiple substructures and configurations. In addition, they have different vapor pressures so that a type may be present in high concentration but have a lower vapor pressure than one present at lower concentration. As a consequence, it is notoriously difficult to analyze the relations between odor molecule types and the perceptions to which they give rise.

Box 7.3 provides an introduction to how wine aroma molecules vary in their basic skeletons. It is obvious from this brief survey that the multidimensional world of aroma molecules is not a simple one. The traditional goal of enology is to understand the relation between the molecular structures and the perceptions they arouse. A goal of neuroenology is to contribute to this effort by understanding how the odor molecules interact with odor molecule receptors and how this is encoded in brain circuits. This should help to build a better foundation for understanding how the brain creates coherent flavor perceptions from these apparently chaotic elements.

BOX 7.3
Representative Wine Aroma Molecules

Isovaleric acid   Isoamyl acetate   Cassis

Ethyl valerate   L   Linalool   R   Ethyl acetate

Isovaleric acid: The aroma is cheesy, sweaty, foot-like, leathery, or barnyard. As a volatile ester, however, it is like a perfume, smoky, or spicy; these aromas arise as a result of the action of *Brettanomyces* yeast.

Isoamyl acetate: The aroma is like juicy fruit, banana, or pear.

Cassis: The aroma is black currant and fruity.

Ethyl valerate (ethyl pentanoate): The aroma is pleasant, with a fruity apple flavor.

Linalool: The aroma is sweet, floral (with a touch of spice), and petitgrain-like, while the aroma of the (*R*)-form is more woody and like lavender, coriander, and sweet basil.

Ethyl acetate: The most abundant ester in wine. In young wines at low concentration, it contributes to fruity flavor, while in older wines at higher concentration, it resembles vinegar.

Research provides evidence that the brain creates a different image of each type of molecule, as explained in the text.

Stimulating the Smell Receptors

Why is it important to know something about the molecular basis of wine aroma? How wine gives rise to our aroma perception has been a mystery through the ages. In the modern age, researchers realized that this involved smell molecules stimulating molecular receptors in the sense cells in the nose. In the 1980s, the mechanisms for signal molecules stimulating their receptor molecules elsewhere in the body were becoming known. The mystery of smell began to unravel when one of my former students, Doron Lancet from Israel, identified a second messenger pathway in the olfactory cells that was similar to the pathway in the photoreceptors in the eye. Second messengers are molecules that receptors activate to act on different target molecules in a cell. The researchers hypothesized that the olfactory receptor might have a molecular basis similar to the photoreceptor. They also hypothesized that to be able to respond to many different odor molecules there would likely be many different receptor types—not just three as in vision but perhaps 1,000 or more.

The key step was taken by a young neuroscientist, Linda Buck, working in the laboratory of a leading figure in molecular biology, Richard Axel, at Columbia University in New York City. She had been studying various systems, including how a simple creature like the garden slug produces hormones, but became intrigued by the sense of smell and threw herself into the search for the receptor. Her laboratory experiments in rodents produced a sensational result: she confirmed the prediction that smell receptors are similar to photoreceptors and even more intriguing, that they are encoded by a large gene family of over 1,000 members, the largest in the entire genome. The importance of this finding, not only for olfaction but also for biology in general, was immediately recognized, leading to Buck and Axel receiving the Nobel Prize in Physiology or Medicine in 2004.

We and others began to study how odor molecules activate these receptors. The receptor molecule snakes back and forth across the membrane seven times to form a kind of barrel around a central opening, similar to the receptor molecules in photoreceptors but specialized for responding to odor molecules. A Yale student in our lab, Michael Singer, postulated with a molecular model that the odor molecules enter a binding pocket in the barrel to activate it, which in turn leads to activation of the previously discovered second messenger pathway to generate an impulse response in the receptor cell. Another of my former students, Stuart Firestein, with Haixing Zhao, a graduate student of John Carlson at Yale (showing how science benefits from combined expertise), demonstrated that a particular receptor called *I7* could be activated best by an eight-carbon aldehyde

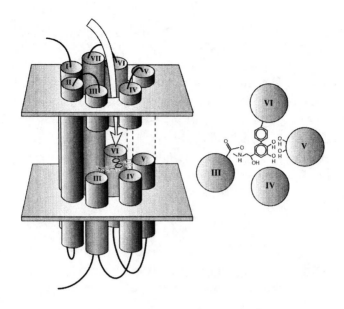

FIGURE 7.2

How an odor molecule interacts with an olfactory receptor molecule: (*A*) the receptor seen from the side, within the two membrane layers, showing the seven transmembrane segments forming a central opening that the odor molecule enters (arrow); (*B*) the odor molecule interacts inside the binding pocket with parts of the molecules that compose several of the sides of the receptor barrel. The different "determinants" on the odor molecule and the different molecules in the barrel segments determine how strongly or weakly this odor molecule activates this receptor. (Redrawn from G. M. Shepherd. Discrimination of molecular signals by the olfactory system. *Neuron* 13 [1994]: 771–790)

molecule and to lesser degrees by shorter or longer aldehydes on either side. Our model of I7 and the different stimulating molecules supported the idea of a binding pocket where this interaction might take place. Figure 7.2 shows simplified views of the receptor. The seven segments are in parallel, forming a barrel; inside is a binding pocket with various molecules sticking into the pocket, where they interact with the odor molecule—in this case the eight-carbon molecule octylaldehyde (*octanal*).

Since that time, a number of experiments and models have greatly extended this concept. We suggest that the complicated structure of the odor molecule bristles with many "determinants," which interact to varying degrees with many binding sites within the receptor. The binding sites can thus interact to different degrees with many different determinants on different molecules. These form the "primitives" of the olfactory sense. No wonder smell is complicated!

And just to make things a bit more complicated, a former student, Minghong Ma, and her colleagues at the University of Pennsylvania, have found that some olfactory receptor cells can respond to touch as well as odor molecules. The strength of a sniff may thus be combined with the strength of activation by specific odor molecules in the signal sent by the sensory neurons to the brain. Ma has suggested that this mechanical sensitivity may help to synchronize the smell responses with breathing and sniffing, enhancing smell acuity. Repeated light sniffing, as described in chapter 9, may enhance this effect.

Adaptation

A factor that comes into play in any test with smell is adaptation, defined as the decrease in the amount of a sensory response with repeated stimulation. Several terms are used for this phenomenon, as summarized in box 7.4.

All these factors may be occurring during the continued drinking of a wine. They mean that a wine expert has to pace himself or herself in testing a wine or a series of wines. Most wine professionals limit their testing to 20 to 30 wines during a day's work, although Robert Parker has been known to sample as many as 200 or more. This may reflect an ability to recognize cues using semantic devices, an ability we will discuss further in chapter 18. Much more on dealing with the effects of adaptation will be found in most books on wine tasting.

BOX 7.4
Different Terms for the Decrease in a Response with Continued or Repeated Stimulation

Adaptation	Decrease due to physiological factors
Desensitization	Decrease due to buildup of biochemical factors
Fatigue	Decrease due to fatigue from buildup of response products

How Good Is the Human Sense of Smell?

If this is how an individual receptor molecule is activated, how does the activation of all the receptors give rise to our perception of the aroma of a wine? To answer that question, we first need to ask: How good is the human sense of smell? There is a widespread belief that it degenerated during evolution and is quite limited. It is understandably not a belief that most wine experts have, but what is the evidence?

The receptor molecules are located in *cilia*, fine hairs extending from the cells into the overlying mucus, where they are stimulated by the odor molecules. The first hypothesis is that the more receptor cells there are, the better the sense of smell. Best estimates are about 10 million receptor cells in humans, which seems more than enough for sensing the flavors of the food we eat and the aromas of wine. However, a rodent has up to 20 million and a dog up to 100 million, which suggests they have more sensitive noses. However, these larger numbers may be needed for special smell-driven behaviors that these and other animals engage in.

A second hypothesis is that more receptor types mean a better sense of smell. Molecular biological studies show that rodents have the most functional gene types—around 1,000, compared with fewer than 400 in humans. This seems to clinch the case that humans have a limited sense of smell. But dogs, with their huge number of receptor cells, have only around 800 receptor types, scarcely twice the number of humans.

At present, the relation between receptor types and smell acuity is not clear. We need to know more about the rest of the smell pathway in the brain to answer this question.

CHAPTER EIGHT

Orthonasal Smell
Creating a Wine Aroma Image

We have seen how the information contained in odor molecules is transferred to the receptor cells to form the basic information units, the "primitives," of the sense of smell. How does the brain build the perception of a wine aroma from this beginning?

Creating Representations of Wine Molecules

Rome was not built in a day, and the perception of smell also takes many steps, as summarized in figure 8.1. In chapter 7, we discussed how determinants on the smell molecules take the first step by activating different smell receptor molecules in the receptor cells. In this chapter, we show how a series of brain circuits process the information as the basis for the perception of wine flavor.

We begin by asking: How does the brain encode information about its sensory worlds? The general principle is to represent sensory stimuli as spatial patterns called *sensory images*. We have seen that this is easy to understand in vision, where the eyes represent a spatial scene as an image in the retina that is transferred in the visual pathway all the way up to the visual cortex in the posterior part of our brain. Similarly, the sense of touch is represented as a map of the body in the somatosensory part of the cortex, with larger areas for the lips and mouth, reflecting their heightened sensitivity. Even the auditory system represents different frequencies of sound by a frequency map in the auditory cortex. And

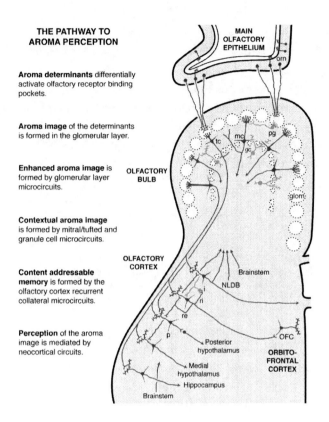

FIGURE 8.1

Steps in wine aroma processing. The steps are explained in the text. (Adapted from G. M. Shepherd, *Neurogastronomy: How the Brain Creates Flavor and Why It Matters* [New York: Columbia University Press, 2012])

there are many other kinds of spatial patterns in the brain, including *motor* maps for the outflow of movement commands and even *images of desire* for food cravings and addictive drugs (chapter 19).

What about smell? Do smell molecules act in the nose in spatial patterns that could represent different smells? In other animals, the nasal cavity contains highly convoluted passageways that have many functions: cleaning, warming, and moistening the inhaled air, as well as generating turbulence to enhance stimulation of the olfactory receptor cells. By comparison, the human nasal cavity is much simpler: the inflow pathway is relatively straight, and there is just enough turbulence to stimulate the receptor cells at the top of the cavity. We saw evidence that molecules in the inspired air may be absorbed in the mucus to different extents, forming a pattern of activation of the receptor cells based on the physical properties of absorption. However, these properties seem insufficient to encode thousands of different molecules emanating from thousands of different wines. It seems more likely that encoding takes place through specificity in the responses of the molecular receptors to the odor molecules and through an organization of the smell pathway that enhances that specificity.

Creating the Aroma Image

A key lies in how the receptor cells send their signals to the next stage in the pathway, the *olfactory bulb*. The olfactory bulb is similar to the retina in being the first station for processing the sensory responses. Research has shown that each olfactory receptor cell expresses only one of the receptor types, and all of these cells send their fibers to one site, one module, in the olfactory bulb, as indicated in figure 8.2. This module is called a *glomerulus* (from the Latin *glome*, a ball containing many small elements). The small elements, in this case, include the endings of several thousand fibers, all carrying the signals from one type of receptor. The advantage of this arrangement is that the smell world is noisy, with many volatile molecules bombarding the olfactory receptors. This is particularly true of wine aromas, which may contain hundreds of different types of volatile molecules. Having all cells expressing a given receptor connect to one module raises the "signal-to-noise ratio" at the module, enhancing the detection of the molecules that activate that receptor.

Each receptor subset connects to two glomeruli in the olfactory bulb so that in most animals receptor subsets expressing up to 1,000 receptors connect to up to 2,000 glomeruli. For any sniff, the odors differentially activate the different receptors to set up a pattern of activated glomeruli. The pattern elicited by a six-carbon aldehyde, which has a grassy smell, is shown in figure 8.2A. In figure 8.2B, the pattern is shown as originally recorded (*left*), compared with the patterns of aldehydes with five and six carbon atoms. The patterns can be seen to overlap but are different enough to tell apart. A mouse can also tell them apart by smell,

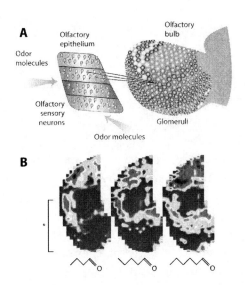

FIGURE 8.2

How odor images are formed in the olfactory bulb: (*A*) olfactory sensory neurons expressing a given olfactory receptor send their outputs to one module (glomerulus) in the olfactory bulb. The different modules are activated to different extents, forming an activity pattern (smell image); (*B*) the activity patterns for three odor molecules, differing by only a single carbon atom and its two hydrogen atoms. The patterns are overlapping but different. (From G. M. Shepherd, Smell images and the flavour system in the human brain, *Nature* 444 [2006]: 316–321)

as psychobiologist Matthias Laska and student Dipa Joshi, working in our lab, showed in behavioral experiments.

These patterns created in the brain are similar to the internal images in other systems we noted previously. We therefore call these patterns *odor images* or *smell images*. We hypothesize that they are the spatial representation of the stimulating molecules, created entirely by the circuits within this region of the brain. They are how the brain uses its internal space to represent these nonspatial objects, analogous to the way sound frequencies are represented in spatial maps in the auditory system. The patterns themselves are not consciously perceived but rather are the first step toward the construction of mental images of smell and their conscious perception (chapter 9).

The detailed odor images shown in the figure are functional MRI patterns obtained with very strong magnets (up to 9.6 tesla) in anesthetized laboratory animals. These methods, and several others that have demonstrated the patterns, are not possible to use on humans. However, the basic circuit organization is similar, and it seems safe to hypothesize that stimulation with the aromas of wine gives complicated patterns in humans similar to those in animals—even more so in awake, behaving humans eager to test their wine-tasting abilities through wine aroma images!

The brain creates these spatial patterns; they are not in the stimulating molecules themselves. It is a new idea not found in books on wine tasting, with one exception. In a 1999 article in a special issue on wine tasting in the *Journal International des Sciences de la Vigne et du Vin*, Frédéric Brochet (who later took part in the study of the effect of wine color on perception+) reproduced a microphotograph of the olfactory bulb responding to an odor, with the caption: "On the surface of the olfactory bulbs, the activation of the glomeruli draws an activation map which is specific to each odor." This statement implied that this is the first step in the formation of mental images of wine perception. It supported our original demonstrations in 1975 and 1979 and to my knowledge is the first explicit application to wine tasting.

Enhancing the Wine Aroma Image by Lateral Inhibition

We have seen that in vision the elements of a scene are recombined into a neural representation, with certain elements enhanced (edges, motion, color) and others reduced (unchanging background). A main mechanism for this is lateral inhibition, which as we have mentioned is very simple: more active cells suppress less active cells. It creates a high-contrast image similar to a photo in stark black and white.

As discussed in chapter 7, wine, like so many natural odors, contains dozens or hundreds of odor-stimulating molecules. They constitute a kind of aroma "scene" analogous to a visual scene. The varying strengths of activation of the different receptors are encoded in the varying strengths of responses of the olfactory glomeruli. We can predict in analogy with vision that the brain would use lateral inhibition to enhance the responses to some molecules and inhibit the responses to others.

How might this work in smell? My graduate thesis project produced evidence for very strong lateral inhibition in the olfactory bulb. Subsequently, a team at the National Institutes of Health led by Wilfrid Rall—including Thomas Reese, Milton Brightman, and myself—discovered a microcircuit between granule cells and mitral cells that appeared to mediate this lateral inhibition. We predicted that it would play an essential role in processing smell.

Current research is indicating that lateral inhibition is used for several distinct operations in the olfactory bulb. At the first level, in the glomerular layer, modeling studies by Christiane Linster and Thomas Cleland at Cornell University show that more active glomeruli can inhibit less active glomeruli to carry out contrast enhancement on the aroma image. Glomerular layer interactions are also used to adjust the patterns due to different odor concentrations. The contrast-enhanced pattern is sent to the second level, where a model with Michele Migliore, Francesco Cavarretta, and Michael Hines at Yale and Palermo, Italy, suggests that lateral inhibition by the granule cells synchronizes the patterns for output to the olfactory cortex. In summary, we see that a sequence of distinct operations, employing lateral inhibition in different ways, is needed to begin to create the brain's enhanced representation of wine aromas. In figure 8.1 we call this a "contextual aroma image" to reflect the fact that it is a synchronized image binding together all of the aroma "scene."

What is the evidence for contrast-enhanced aroma images among the many hundreds of possible aromas? A strategy was introduced by a former student, Kensaku Mori, then in Osaka, Japan. He decided to simplify the problem by studying responses to a chemical series of odor molecules that varied along only one dimension. He and his colleagues Masayuki Yokoi and Shigetada Nakanishi began with straight-chain aldehyde molecules varying from 3 to 10 carbon atoms, as those shown in figure 8.2. They are called aldehydes because they have what is called a *functional group*, an oxygen molecule attached by a double bond to the last carbon, at one end of the chain. As noted, the functional group gives many of the members of this series a grassy smell. These kinds of aldehydes are present at low levels in many wines and serve as an example of the kind of analysis one would like to do with more prominent aroma compounds.

Mori and his colleagues recorded from individual output neurons, the mitral cells, stimulated with single molecules of increasing length, to test the hypothesis that individual mitral cells would be preferentially tuned to different members of the series. Their recordings verified this hypothesis; they found that most cells responded best to one of the members and responded more weakly to neighboring members in the series, reflecting differing sensitivities of the receptors in the nose. In addition, to their surprise, they found that a given cell responding best to one molecule inhibited responses to neighboring molecules. It appeared to be similar to the visual system, expressing the general principle of lateral inhibition that a cell preferentially tuned to a given stimulus inhibits cells less tuned to that stimulus. In aggregate, this means that different cells are constantly competing to give the strongest responses to their particular part of the aroma scene and inhibit the others.

We postulate that in orthonasal (as well as retronasal) sensing of a wine, we are discriminating different elements in the aroma by using the detection of types and classes of aroma molecules together with lateral inhibition to enhance them. During evolution, this was presumably a mechanism to identify molecules in foods (insects, fish, plants, fruit, meat) that had smells indicating, for example, ripe fruit and good nutrition. Paradoxically, we do not drink wine for its nutritive value but for pleasure. It appears therefore that in wine tasting a sensory discrimination system for a healthy diet is taken over by a discrimination system devoted to pleasure.

Modulating the Wine Aroma Image by Behavioral States

Processing in any sensory system takes place despite wide ranges in activity and mood, which are called our *behavioral states*. Much of the processing continues whether we are happy, sad, elated, or depressed. In vision, for example, the retina processes the sight of food and wine regardless of these swings in state. This is understandable because few fibers connect the brain to the retina to modulate its operations.

In sharp contrast, the brain is very interested in what is coming in through the olfactory pathway and modulates the olfactory system beginning with the first stages of formation of the odor images in the olfactory bulb. Modulatory fibers originating within the brain are called *centrifugal fibers* because they are directed outward from the brain to the periphery. While most research on olfactory processing focuses on the circuits that process the odor signals entering from the olfactory nerves, the olfactory bulb also receives an extremely heavy input of centrifugal fibers. Fibers come back from the olfactory cortex, to which the olfactory bulb projects. Fibers come from the forebrain, signaling with a transmitter called *acetylcholine* what is happening in our cognitive world. And they come from deep in the midbrain, signaling with transmitters called *noradrenaline* and *serotonin* our behavioral state relative to feeding, alertness, and mood.

The fibers related to feeding are especially interesting for neuroenology. In 1972, an insightful experiment was performed by physiologist Jeanne Pager and a team led by Andre Holley and Jacques Le Magnen, a French physiologist. Pager recorded the impulse responses of mitral cells to orthonasal food aromas in rats that were either hungry or had eaten to satiety. She found that the food responses varied dramatically; they were strong when the animals were hungry and **weak** when they were sated (and could eat no more). This showed that the central behavioral states of hunger or feeling full closely modulate the olfactory bulb circuits.

In summary, processing smell involves dedicated circuits that can be modulated from multiple brain sites reflecting different behavioral states. The brain is interested in everything we smell and attempts to regulate it if it can. For wine tasting, this means that the smell component, both orthonasal and retronasal, is especially sensitive to the behavioral state. So if you are serious about discriminating wines, don't eat too much, and be sure to get yourself into a good mood!

CHAPTER NINE

Orthonasal Smell
From Odor Image to Aroma Perception

The aroma "scene" in the olfactory bulb is the brain's way of representing the stimulating molecules. Its function is to encode the properties of the molecules.

The brain must next convert this fractured pattern into a pattern representing the entire mixture of molecules. This is analogous to taking the "primitives," the fundamental elements of a visual scene (lines, angles, corners), and reintegrating them into a neural representation of the visual scene. For this, the brain needs to transform the scene from the language of the senses to the language of the brain. In vision, this means creating a *gestalt*, a pattern that is recognized as a whole. This is one of the things the brain does best. We learn to recognize objects instantly—not just classes of objects (a human figure at a distance) but even individuals (by a glance at their shape or movement).

For smell, this next step occurs when the enhanced aroma scene in the olfactory bulb is sent to the olfactory cortex. The olfactory cortex has a simple arrangement, as was indicated in figure 8.1 and is shown in more detail in figure 9.1(*A*). The main cell type is called a *pyramidal cell* because its cell body is in the form of a pyramid. From the top of the pyramid, a long fiber (an *apical dendrite*) extends, branching off in several directions before it reaches the surface. The long output fibers from the mitral and smaller tufted cells connect to these branches, activating them when they discharge an impulse in response to the input to the olfactory glomeruli. But instead of a one-to-one connection from a mitral cell to a pyramidal cell, each mitral cell fiber makes connections with

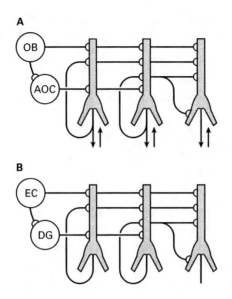

FIGURE 9.1

The olfactory cortical microcircuit functions as a content addressable memory: (*A*) Olfactory cortex: a single fiber from the olfactory bulb (OB) contacts many pyramidal cells at the ends of their main branches (dendrites), and the pyramidal cells in turn send collaterals of their own output fibers to many further pyramidal cells. This all-to-all architecture enables the pattern to be retrievable by every cell, known as a content addressable memory system, essential for memory networks; (*B*) Hippocampal formation: the organization is similar in the hippocampus, which is well known as a memory center. Abbreviations: OB, olfactory bulb; AOC, anterior olfactory cortex; EC, entorhinal cortex; DG, dentate gyrus. (Adapted from K. R. Neville and L. B. Haberly, Olfactory cortex, in *The Synaptic Organization of the Brain*, 5th ed., ed. G. M. Shepherd [New York: Oxford University Press, 2004], 415–454)

many pyramidal cell dendrites. Conversely, each pyramidal cell receives connections from many mitral cell fibers.

Building a Pattern Recognizer

If you are an engineer or a computer scientist, you recognize that this creates a new kind of machine. First, each pyramidal cell receives a little input from every part of the olfactory bulb. My first graduate student, Lewis Haberly, now at the University of Wisconsin, had the first insight into this arrangement in the olfactory cortex. He called it a "content addressable memory," meaning that each site in the olfactory cortex can have its content addressable in order to retrieve the memory of the whole scene. In other words, it is a pattern recognizer. The whole pattern, the gestalt, can be re-created from a small part. The best-known example of this property is the face area of the cerebral cortex. We can see only a small part of someone's face but are able to fill in the rest of the face and recognize it.

A second feature of this kind of device is that, as shown in figure 9.1(A), it needs to have long association fibers making connections from any one cell to many other cells throughout the cortex. These have two actions. First, they provide for the reexcitation of other pyramidal cells to reinforce their responses and bind them into new units for the recognition of different patterns. Second, through lateral inhibition they provide contrast enhancement of the resulting aroma representation. This is shown in figure 9.1(A) for the olfactory cortex along with, for comparison, the hippocampus (see figure 9.1[B]), which is well known to be critical to learning, storing, and retrieving memories. We will discuss the role of learning and memory in wine tasting in chapter 17.

Analytic or Synthetic

Wine tasters are fighting a special limitation on complex odor perception: Is it analytic or synthetic? Donald Wilson at New York University and Richard Stevenson at McQuarrie University in Australia have summarized recent work in their book *Learning to Smell: Olfactory Perception from Neurobiology to Behavior*. This is an old issue in sensory perception. *Analytic* is when you can discern separate components in a mixture. Taste is a good example: you can identify both sweet and sour in the sweet-and-sour sauce so important in Chinese cuisine. *Synthetic* is when the components of the mixture merge into a new whole; smell is usually regarded as this kind of perception. Wilson and Stevenson make a good case that olfactory processing starts out in the olfactory bulb as analytic as it separates the different features of the odor molecules, and then becomes synthetic as it reorganizes them in the olfactory cortex. They analyze how many features can be identified in a complex smell and arrive at the following conclusion:

The more complex the odorant mixture is, and the more overlapping in features and corresponding patterns of olfactory receptor activation the components are, the more difficult task of perceptual grouping facing the piriform (olfactory) cortex. We hypothesize that it is the limit in perceptual grouping or pattern recognition abilities of the piriform cortex that places the upper bound on mixture analysis at three components. Beyond that limit, individual component analyses become faulty and odorant mixtures are processed as a single perceptual gestalt.

Their analysis thus is relevant to professional wine tasting, where the aim is to identify as many components as possible to reveal the origin and quality of a particular wine. As we will discuss in chapter 17, the research cited appears to put a maximum of three components as a limit. This obviously has interesting implications for both amateur and professional wine tasters.

Where Does the Aroma Object Become Conscious?

The olfactory cortex is a simple cortex with only one layer of output cells (see figure 8.1), yet it is sufficient for creating the "smell object," and it plays a major part in the mechanisms underlying smell perception.

The last step in creating the perception of the smell object is making it conscious. Where does the brain "read" the sensory representation it has created and make it into a conscious perception? This is a profound question that neuroscientists and psychologists ask of every sensory system. In other sensory systems, the pathway has to pass through the thalamus, the nerve cells that serve as the gateway to the neocortex.

The smell pathway stands out as being radically different, for several reasons. First, the smell pathway has only a small "indirect" relay from the olfactory cortex through the thalamus (the mediodorsal thalamus, to be exact) to the neocortex. Smell therefore is believed to be unique among sensory systems in not requiring a major relay through the thalamus for conscious perception to occur.

Second, the smell pathway is part of the *limbic system*, the collection of centers just below the neocortex where subconscious centers for

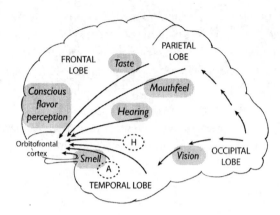

FIGURE 9.2

Overview of the human brain, showing the direct input to the orbitofrontal cortex by the smell pathway and the convergence of relays from the other sensory systems to give rise to conscious multisensory flavor perception.

emotion and memory are found. Smell has direct access to those centers as well. This is presumably one reason why most people remember smells so vividly.

Third, both the direct and indirect pathways arrive at a primary smell cortical area at the front of the brain in the *orbitofrontal cortex* (OFC): "orbital" because it lies just over the orbits of the eyes and "frontal" because it is at the extreme front of the brain (figure 9.2). This is most likely the level at which conscious smell perception occurs. In sharp contrast, the other senses—vision, touch, audition, and taste—have their primary sensory areas further back in the cortex and so require relays to the front of the brain for conscious perception to occur.

By contrast, the smell pathway goes there directly. This presumably reflects the importance of orthonasal smell in reacting to prey or predator smells and of retronasal smell in evaluating the food that is eaten. This brain region is also involved in higher cognitive functions such as making judgments and decisions and planning for the future, so it is prime real estate in the cerebral cortex, and smell is at the center.

The consciousness of smell is therefore special and differs from the other senses. Research on what these differences are and where they occur is only beginning. For example, if the olfactory pathway doesn't pass through the thalamus, how is it coordinated with the other sensory systems that do? Kensaku Mori and his colleagues Masayuki Yokoi and Shigetada Nakanishi addressed this problem by making recordings from cells in the rat piriform cortex and correlating them with the state of the rest of the cortex, whether it was sleeping or waking. They found that the cells responded briskly to odor stimulation during the waking state but only weakly during sleeping states. The weak responses could be changed to strong responses if the brainstem reticular activating them was stimulated to activate the cortex, as it does during waking. The authors concluded that "state-dependent sensory gating in the olfactory system is in synchrony with other sensory systems" through the coordinating actions of the brainstem activating system.

Higher Olfactory Processing Is Complex

We have focused on the initial steps in processing the olfactory input and now move to the central mechanisms for creating perception. This requires a much broader view of the connections involved. For this we can build on the representations of the movements of wine and aroma discussed in part 1 to include the brain systems that create the sensory perceptions. Figure 9.3 begins with the olfactory systems. Starting at the bottom, it shows a hypothesis for the sequence of connections from the olfactory receptor neurons in the nose to the olfactory bulb and to the olfactory cortex. Adapting a schema suggested by Edmund Rolls and his colleagues at Oxford University, we may place the olfactory cortex at Level 1 in the processing sequence. It is concerned with identifying what the stimulus is; for example, what kind of floral or berry flavors does the wine have?

The olfactory cortex projects to the OFC, where the value of the stimulus is assessed in Level 2. This is the primary neocortical area for olfactory processing; as we have seen, its cells have flexibility in responding to a given smell as rewarding or not. By dispatching its output to multiple sites, the OFC is the central coordinating center for the cognitive network that creates the ultimate wine flavor image.

To fulfill that coordinating role, its output is distributed to Level 3, consisting of multiple sites with multiple complementary functions, each carrying out a function with respect to the aroma perception. One site is the *striatum*, whose activity generates habitual behavior; in wine tasting,

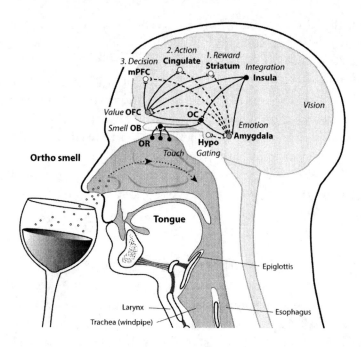

FIGURE 9.3

Simplified diagram of the olfactory pathway to and within the cerebral cortex for conscious perception of sniffing the wine aroma. The pathway has three levels of processing, as explained in the text. The actions (1, 2, 3) resulting from the processing are shown at the top level. Many of these sites are involved in other actions, depending on the behavioral context. (Adapted from E. Rolls, Taste, olfactory, and food reward value processing in the brain, *Progress in Neurobiology* 127–128 [2015]: 64–90; and other authors)

this would reflect training and experience. A second is the *cingulate gyrus*, which is more concerned with selective actions. We might call it actual expertise gained by training. A third is the *medial prefrontal cortex*, whose actions are related mainly to decisions about pleasant aromas. This crucial role of pleasure is discussed fully in chapter 19. Finally, there are connections to the hypothalamus, which is a master node for coordinating hormonal behavioral states with selective actions toward foods and fluids in relation to hunger or satiety. As shown in the diagram, the amygdala acts in parallel with most of these actions, adding an emotional dimension to each connection.

The top levels of processing presumably create what we call the conscious perception of the wine aroma. How might this arise?

Constructing a Conscious Wine Aroma Mental Image

We may be conscious of a smell perception, but in what form is the perception? In vision, for example, a conscious perception is in the form of a spatial image, which is present from the moment the light strikes the retina. An essential aspect of this image is that it is present when we are looking, and it remains if one looks away or remembers it later. It is a mental image of the visual scene.

Is there a corresponding mental image of a smell? Arguing in favor is the fact that a spatial image of a smell is formed in the olfactory bulb and projected onto the olfactory cortex. Although no one has yet seen a representation of smell in the olfactory cortex or in the areas of projection in the neocortex, there is increasing evidence that olfactory perception does involve a mental image. The evidence is too extensive to cover here, so we shall cite a review by Artin Arshamian and Maria Larsson at the Gosta Ekman Laboratory at Stockholm University. It is summarized in box 9.1. They compared research on mental images in vision and olfaction and found many similarities suggesting that there is a mental image of smell, even though we are not conscious of it. The authors note that the ability to form mental images is weaker for smell, and there is more individual variation. Those with more experience and who possess a richer vocabulary for describing odors are best at olfactory imagery.

The take-home message is that while humans are not conscious of a smell's mental image, the perception nonetheless shares many of the attributes of the well-studied mental images of vision and hearing. The remarkable difference is that although one consciously perceives a smell, a smell that has properties of an image, that image nonetheless is not conscious; it cannot be described as an image. This would seem to contrast sharply with vision and hearing. However, it was noted in *Neurogastronomy* that nonrepresentational art, both visual and musical, is also extremely difficult to describe in words. We are very good at recognizing complex patterns, such as someone's face, but lack the vocabulary to describe them. This is exactly the challenge in describing the qualities of a wine aroma and, as we shall see in chapters 14 and 15, a wine flavor.

BOX 9.1
Similarities Between Visual and Olfactory Imagery

1. Perceived and imaginary stimuli have similar properties.
2. Mental images interfere with perceived images.
3. Both show mental hallucinations.
4. Visual and smell dreams have similar characteristics.
5. Vividness of imagined pictures and smells is similar, more vivid in females.
6. Mental images can be accompanied by eye movements or by sniffing.
7. Mental imagery activates primary and secondary visual and olfactory cortical areas.
8. Visual imagery can be elicited by stimulation of the subthalamus, and olfactory imagery by stimulation of the olfactory bulb.
9. Image recall is enhanced by expertise.
10. Functional brain reorganization is stimulated by visual or smell experience.

From A. Arshamian and M. Larsson, Same same but different: The case of olfactory imagery, *Frontiers in Psychology* 5 (2014): 34.

Perception and Cognition: The Role of Learning

The OFC is the first stage in the smell pathway where the strength of the behavioral response can depend not only on the concentration of the stimulus but also on whether it is preferred and whether the preferment is learned. One such ability is *reversal learning*. Edmund Rolls in 1996 was a pioneer in this field. In these experiments, an animal is trained to receive a reward, such as juice, for preferring the smell of amyl acetate over ceneole, and then the reward is switched. Behaviorally, the animal can quickly learn to switch its preferment to get the reward. Experiments show that this behavioral switch is correlated with a switch in the activity of individual cells in the smell area of the OFC. This switch is not seen in the olfactory bulb or the olfactory cortex at lower levels in the pathway. Some cells do not show reversal; this is interpreted to reflect the stability of the relation between an odor and a particular behavioral response maintained by some neurons and their circuits.

The implication for wine tasting is that the plasticity of synapses at the level of the OFC enables one not only to learn to better detect and discriminate specific aromas but also to change a preference from one aroma to another. This can help explain the change for a beginner who starts by preferring a wine with a simple fruity aroma but, with experience, develops a preference for a wine with more subtle aromas. It can also allow for the gradual accumulation of experience to lead to a preference for new types of varietals and a new appreciation for the qualities of older wines. For these behavioral changes, the OFC appears to be critical.

There is much work in neuroscience on the property of plasticity, that is, the ability of a synaptic connection to have its strength molded by experience. This work is summarized in Wilson and Stevenson's book *Learning to Smell*.

In addition to plasticity in learning, the OFC is critical for integrating smell with other senses, by the convergence shown in figure 9.2. Taste, of course, is the sense most closely associated with smell. We discuss this integration in chapter 15.

For orthonasal testing, the most relevant integration is with touch. In the next chapter, we will discuss touch receptors in the mouth that food and wine activate. The nose also contains touch receptors that are stimulated by the air (in addition to the mechanical sensitivity of the olfactory receptors themselves). Their responses give rise to impulses that travel up to the touch area of the cortex and are relayed forward to the smell area, where they are integrated by cells that fire in response to both the odor stimulus and the touch stimulus. The touch sense is so delicate that many responses we think are to smell are in fact to stimulation from the air we sniff in. This can be especially true of pungent smells. Other cells in the OFC have been shown to be activated by smell and visual stimulation. These experiments support the evidence that the OFC is a key site in wine tasting, where all the senses are combined to give a conscious perception of the unified wine object image.

CHAPTER TEN

Touch and the Mouthfeel of Wine

We have noted that when we take our first sip of a wine, the touch of the wine to our lips fixes the wine to our mouths, and our brains from then on interpret all of the sensations of the wine as coming from our mouths.

Our detection of the presence of the wine in our mouths is often called touch or feel or tactile sense; scientifically, it is called *somatosensation*; that is, all the sensations that come from the *soma*, meaning "body." It includes all the sensations that arise from sensory receptors in the skin or, in this case, the mucus membranes of the mouth. Although we may say that much or most of a wine's flavor is due to retronasal smell, this is mostly relevant to the sensations of smell that dominate the flavor. Often subconsciously, stimulation by the touch receptors is a major factor in whether we like the wine. According to Jackson, the average wine consumer places more importance on the mouthfeel of a wine than on its fragrance.

It is a fiendishly complicated system, reflecting the fact that the skin is our interface with the outside world, and the mucus membranes are our interface with anything coming into the body from the outside world. Both need to have an array of sensors to detect the ways that the outside world impinges on our bodies. The serious wine student will discover that being aware of the structure of the mucus membranes, their various receptors, and the sensations they produce will enrich the wine-tasting experience.

The Different Sensory Receptors of the Mouth

The mucus membranes differ according to the area of the mouth (figure 10.1). The thinnest membrane is found on the inner lips and cheeks, on the floor of the mouth, under part of the tongue, and on the soft palate. It consists of a deep spongy layer overlain with a continuous tight layer called the *basement membrane*, with two layers of cells on the surface. This thin type of mucus membrane is relatively delicate. It obviously is vulnerable to trauma, as you discover if you accidentally bite your cheek or drink superhot liquids. Elsewhere in the mouth— the gums, the top of the tongue where the taste buds are, and the hard palate above—the mucus membrane has two more layers, topped off with a surface layer of dead cells that are hardened (*keratinized*). It is similar in construction to the skin, with its toughened surface. This kind of mucus membrane is more resistant to trauma but less sensitive to touch.

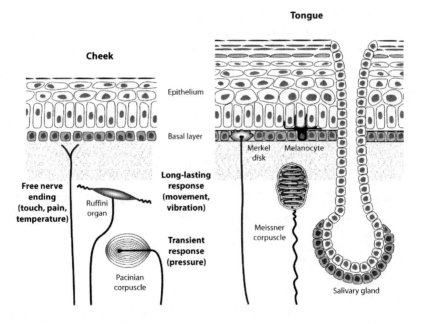

FIGURE 10.1

The mucus membranes and structures in the mouth and the sensory receptors for mouthfeel.

Several types of sensory cells are embedded in these two types of membranes, tuned to different physical aspects of food and liquid.

Light touch. The receptors for light touch sense the intake of the wine and localize where it is swishing about in our mouths. They include the endings of nerve fibers below the spongy layer and special tiny organs that include the Merkel complex, Ruffini organ, and Meissner organ. By their different structures, they are tuned to slightly different touch properties, such as *location, slow or fast movement, direction of movement,* and *force*. While neuroscientists study how each organ relates to each property, the relevance to you as a wine taster is to know that each of us has a set of these sensors and that discrimination can be enhanced by attempting to differentially activate them. Variations in their sensitivity can play a role in our judgments of a wine.

Pressure. The Pacinian corpuscle, a small organ under the mucus membrane, senses pressure. It consists of many tightly bound layers in the form of an onion; the layers are sensitive to pressure but quickly rebound so that the sensor responds only to rapidly applied pressure, like pressing the tongue against the cheek or hard palate.

Temperature. Temperature sensors register a wide range of temperatures in the mouth, from near-boiling water (up to 100°C, or 212°F) down to ice water near 0°C (32°F). Temperature is an important variable to control in wine tasting but within a much narrower range. Reds should be room temperature, around 22°C (70°F), and white wines are drunk cool (around 10°C, or 50°F; chilled wines can go down to 5°C, or 40°F). Temperature is sensed by free nerve endings in the mucus membrane.

Pain. Pain has a role to play in eating food that contains hot peppers, which many people love and many others avoid. However, wines are meant to give us pleasure, not pain, so we can set pain aside as playing a significant role in wine tasting.

Mouthfeel

There is a second class of physical touch stimuli that activates not single sensors but multiple sensors by the dynamic action of the wine, whether by movement, volatility, or combination with saliva. The perceptions these give rise to are extremely important for the wine taste. Collectively, they are often termed *mouthsense, mouthfeel,* or *texture.* Ann Noble, a pioneer in research on wine tasting at the University of California, Davis, introduced the concept of a wheel representing different wine properties. An example is shown in the "Mouth-Feel Wheel" developed by Richard Gawel, A. Oberholster, and I. Leigh Francis (figure 10.2), which has two levels of categories under the main types of touch sensations: Feel and Astringency. This variety of sensations reflects the variety of sensory end organs that are shown in figure 10.1 and the variety of physical properties of the wine.

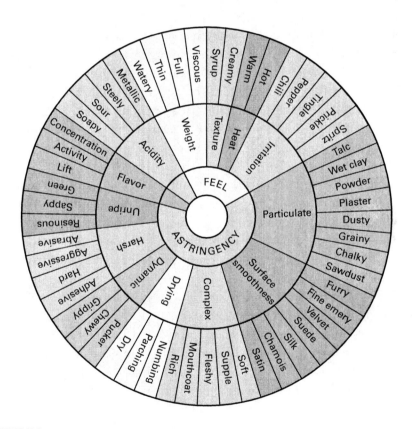

FIGURE 10.2

Red wine mouthfeel wheel. (From R. Gawel, A. Oberholster, and I. L. Francis, A 'Mouth-feel Wheel': Terminology for communicating the mouth-feel characteristics of red wine, *Australian Journal of Grape and Wine Research* 6 [2000]: 203–207)

This helps to identify the main types of mouth sensations. The individual taster can develop his or her own system. As an example, here are terms from Robert Parker's descriptions of Châteauneuf-du-Pape:

Full body
Gobs of glycerin and fat, chewy
Lushness, glycerin, and fatness
Full bodied, chewy mouthful of wine
Soft tannin, full-bodied moderate tannin in the long finish, staggering
    concentration
Rich, fat, and unctuous
Full body, unctuous texture, velvety-textured
Rich, fat wine that is oozing with fruit; glycerin; and alcohol

"Unctuous," by the way, has several meanings. According to *Merriam-Webster*, the first meaning is "fatty, oily; smooth and greasy in texture or appearance," which presumably is what is meant in this context. However, it can also mean "marked by a smug, ingratiating, and false earnestness or spirituality." Only Parker knows if that is also what is meant!

And in a description of Gigondas: "Full bodied, with splendid concentration, moderate tannin, and loads of glycerin." And Burgundy: "White: full-bodied, unctuous, buttery."

These quotations make it clear that mouthsense is critical to wine taste. What are the physical properties of the wine that give rise to these qualities?

Table 10.1 summarizes the physical and molecular composition of most wines that contribute to mouthfeel.

All these elements come from the grape and its fermentation products. Different varietals have different relative amounts of each of these elements. But all must have a balance between them so that one quality does not become too dominant. Judging the balance created by the grape, its fermentation, and the brain is the main challenge to all wine tasters.

TABLE 10.1
Physical and Chemical Elements That Give Rise to Wine Mouthfeel and Texture

Element	Mouthfeel/Texture
Water	"Body," touch
Tannins, Phenols	Astringency
CO_2	Bubbles, prickliness
Ethanol	(High) burning, "body" (dry wines)
Glycerol	"Body" (minimal effect)
Minerals (Cu++)	Metallic (increased by tannins)
Sugar	"Body"
Organic Acids (tartaric, malic, lactic succinic, citric)	Freshness, tartness

Qualities of Mouthsense

How does the brain create the qualities of mouthsense? Here are some examples:

Astringency is particularly important. Tannins taste astringent because they bind with salivary proline-rich proteins and precipitate them. This leads to increased friction between mouth surfaces, and a sense of dryness or roughness, due to the way that molecules in the wine combine with molecules in the saliva and in the surface of the mucus membranes to form conglomerates. The sense of roughness contributes to the perception that the liquid has substance, which is sometimes called the *body* of the wine. Another term for describing it is "chewy"; one feels like it has a body one can chew, a quality also produced by higher levels of alcohol.

The importance of astringency to wine flavor is further explained by Richard Gawel:

Red wine astringency can be assertive and yet display very subtle nuances. Many words have been used to describe these. They include those that have been evoked by the feeling of having fine particles on the surface of your mouth (*Powdery, Chalky, Grainy*), those that relate to the roughness of the feeling inside the mouth (*Silky, Emery, Velvety,*

Furry), and others can be related to the fact that the very astringent wines often cause your mouth to move (*Pucker, Chewy, Grippy and Adhesive*). Other positive terms include *Soft, Supple, Fleshy, and Rich*, and on the unpleasant downside, there are sensations of *coarseness and hardness*.

These and related properties can be located in the wine mouthfeel wheel (see figure 10.2). It is remarkable that these many different sensory properties are produced by the elements of table 10.1, the receptors of figure 10.2, and the circuits in the brain, as we discuss in chapter 12.

It may be possible to measure astringency more exactly. Scientists are developing an optical sensor based on what Joana Guerreiro and her colleagues call a *localized surface plasmon resonance* (LSPR) to study interactions between small molecules and protein. This is a quintessential product of nanotechnology, a harbinger of future developments in neuroenology:

> The sensor is a small plate coated with nanoscale gold particles. On this plate, the researchers simulate what happens in your mouth by first adding some of the proteins contained in your saliva. After this they add the wine. The gold particles on the plate act as nano-optics and make it possible to focus a beam of light . . . so as to precisely measure something that is very small—right down to 20 nanometres (20 billionths of a meter). This makes it possible to study and follow the proteins, and to see what effect the wine has. It is thereby possible to see the extent to which the small molecules have to bind together for the clumping effect on the protein to be set off.

The authors go on to say:

> The sensor expands our understanding of the concept of astringency. The sensation arises because of the interaction between small organic molecules in the wine and proteins in your mouth. This interaction gets the proteins to change their structure and clump together. Until now, the focus has been on the clumping together that takes place fairly late in the process. With the sensor, we've developed a method that mimics the binding and change in the structure of the proteins, i.e. the

early part of the process. It's a more sensitive method, and it reproduces the effect of the astringency better.

This illustrates one of the new methods that is being developed to enable wine to be tested in more quantitative ways. It can be combined with the methods of chemical analysis discussed in chapter 7 to build a more quantitative characterization of the elements of flavor.

By contrast with astringency, "smooth," "velvety," "soft," and "slippery" reflect ingredients in the wine that restrict the interaction of the wine with the mucus membrane so that the wine glides over the surface without adhesion. The different terms reflect not only the amounts of such an ingredient but also the interactions of that ingredient with other ingredients in the wine. High levels of sugar and ethanol make the wine "thick" and "viscous," giving it body. "Fresh" can reflect the aeration of the wine, apart from carbonation. These properties provide a rich keyboard for the harmony within a wine.

In summary, independently of taste and smell, mouthsense fixes our attention on the physical properties of the wine and their interactions with the saliva and the mucus membranes of the mouth. It is a key partner in creating the overall flavor of wine. From the receptors, nerves carry the signals to the brainstem for relay to the thalamus and ultimately to the cortex. The complex network of cortical connections that gives rise to the conscious perception of mouthfeel and combines it with taste is described in chapter 12.

CHAPTER ELEVEN

Taste Modalities and Wine Tasting

When we use the word "taste" to apply to wine, we think we know what we are talking about: the experience of the wine in our mouths. But as we explained at the start of the book, the experience is partly an illusion. It should be no surprise, therefore, that the word itself can be misleading. For some guidance through this terminological thicket, let us start by consulting the experts on words, the Oxford dictionaries. Here you will find a series of definitions of "taste" as a noun that make the problem clear. We adapt them for application to wine (box 11.1).

BOX 11.1
Defining "Taste"

The first example of the meaning of the term "taste" is, "The sensation of flavor perceived in the mouth and throat on contact with a substance": "The wine had a fruity taste." This is the whole experience of the wine. As we have already emphasized, a quality such as fruity is actually due more to retronasal smell than to the taste receptors in our mouths, so the term "flavor" is more appropriate for this use. (Interestingly, English is one of the few languages that has a separate word for the overall taste; it exists in French [*saveur*] but not in German or many other languages.)

A second meaning is the specific sense of taste due to taste receptors in the mouth, which is what this chapter is about.

A third meaning is, "A small portion of food or drink taken as a sample": "Try a taste of cheese." This meaning also applies to wine tasting, which traditionally involves taking small samples of a series of wines to test and compare them.

The meanings up to now have been relevant to the sensory experience of a wine. However, there are further meanings related to our subjective responses that get more problematical.

A fourth meaning is, "A person's liking for particular flavors": "This pudding is too sweet for my taste." This usage combines how something is *perceived* with whether or not one *likes* it. It thus blurs the difference between the perception of a food such as wine and the emotion it arouses. We must be on our guard for making judgments like this.

A fifth meaning is, "A person's tendency to like or be interested in something": "He found the aggressive competitiveness of the profession was not to his taste." We have now passed beyond the uses of taste related to food and wine and are applying the word to many contexts of life in which the perception of what is happening is mixed with the emotion it arouses.

A sixth meaning is, "The ability to discern what is of good quality or of a high aesthetic standard": "She has frightful taste in literature." With wine, this could apply to people just learning how to judge wines or to degree of sophistication.

A seventh meaning is, "Conformity or failure to conform with generally held views concerning what is offensive or acceptable": "That's a joke in very bad taste." This could apply to wine tasting, though it would not be clear whether bad taste was due to bad judgment about the quality of the wine or a controversial preference for a wine that no one else liked.

We could summarize the complicated terminology of taste with the following sentence that combines them all: "The taste of the wine (1) had a sweet taste (2) when I took a taste (3), which was too strong for my taste (4) and therefore not to my taste (5), though I may not have the best taste (6) when compared with your taste (7)."

Oxford Dictionaries: Language Matters, s.v. "taste," http://www.oxforddictionaries.com/definition/english/taste.

Because of these many shades of meaning, we will use the "flavor" of food and wine (the first meaning) interchangeably with the term "taste" when there is a need to clarify. Despite these problems, "taste" will obviously continue to be used with regard to wine, so we will be alert to the sense in which we are using the term. Another advantage of the use of "taste" is that, unlike "flavor," it is also a verb, so it can be useful as both transitive ("She tasted the wine") and intransitive ("The waiter served the wine for him to taste").

Taste as applied to the basic tastes can be referred to as *primary taste*, which is what this chapter is about.

The Five Primary Tastes

We all know that we taste (second meaning) with our tongues and that the taste sensors are in taste buds distributed across the tongue's surface (they are also found elsewhere in the mouth and throat). Because they are accessible to study on the tongue, stimulation with different pure substances (*tastants*) has been possible since ancient times. Aristotle thought there were seven tastes: sweet, salty, sour, bitter, astringent, pungent, and harsh. Through the centuries, sweet, salty, sour, and bitter have persisted while other candidates have been suggested, such as fatty, aromatic, and putrid (by now you can recognize that these reflected mostly the contribution of smell). Nearly all of this work was done in Europe. Around 1900, a Japanese chemist, Kikunae Ikeda, at the Tokyo Imperial University, brought forth evidence for a new taste he called *umami*, for "pleasant savory taste." There was controversy about whether this really was a separate taste until recently, when the receptor for it was discovered, so it is now accepted as one of the five primary tastes.

Each of the first four primary tastes is distinctive in its mechanism and perception; umami is more subtle in its perception. These primary tastes play big roles in cuisine. Sweet, for example, is a primary driver of food consumption, responsible for the widespread cravings for foods and beverages that contribute to obesity and health disorders, such as diabetes. Soft drinks contain enormous amounts of sugar: 7 teaspoons in a 12-ounce can.

TABLE 11.1
The Five Tastes

Primary Stimulus	Primary Receptor	Primary Perception
Sugars	Taste receptor T1R2+T1R3	Sweet
Alkaloids	Taste receptor T2R1+T2R50	Bitter
Protons	Proton channel PKD2K1+TRP	Sour
Salt ions	Salt channel	Salty
Glutamate	Taste receptor T1R1+T1R3	Umami

Wine is very different in this respect, where balance is the key. All of the primary tastes, so prominent in our cravings for food, are muted in wine. However, each makes its contribution.

As in the case of color, the tastes of sweet, salty, sour, bitter, and umami are *perceptions* the brain creates from stimuli in the *physical* world. Similar to the way that color perception arises from different frequencies of light, we need to explain how taste perceptions are created from physical properties of the primary stimuli. How this occurs starts with explaining how receptors on the tongue detect these physical properties. The five primary taste stimuli are summarized in table 11.1, together with their primary receptor protein and primary type of perception. Note that in discussing taste, the stimulus (for example, sugar) must be kept distinct from the primary perception (for example, sweet). Note also that more than one type of stimulus may contribute to a given type of perception.

The Tongue as a Sensory Organ

We have seen that the tongue is at the center of wine tasting, first, as a *motor organ* for moving the wine sip about in the mouth and second, through its role in mouthsense. We now consider its third role, as the main sensory organ for *taste* (second meaning in box 11.1).

The special anatomical unit for taste is the *taste bud,* a tiny pocket of sensory cells located in outcroppings from the tongue's surface called

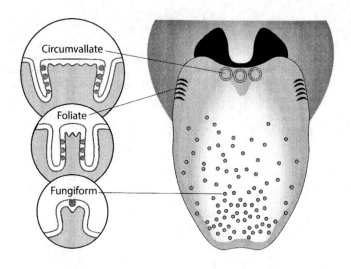

FIGURE 11.1

Taste buds on the tongue. Agitation of the wine in the mouth is needed for the wine to reach all the taste buds distributed over the tongue (and in the throat). (From J. Chandrashekar et al., The receptors and cells for mammalian taste, *Nature* 444 [2006]: 288–294)

papillae (figure 11.1). The papillae are of several types. Some are mushroom-shaped bulges (*fungiform* papillae), some are in folds on the sides of the tongue (*foliate* papillae), and some are in rounded craters (*circumvallate* papillae) at the back of the tongue. Nobody knows why these different types exist, except that it means there are slightly different ways that tasty substances can access the taste buds.

It used to be thought that the papillae carried taste buds specializing in different tastes, forming a "taste map" on the tongue, but lab experiments have disproven a strict map. There may be different sensitivities for each taste, but in general there is considerable overlap of taste buds containing cells responding to the different taste stimuli (see figure 11.1). In fact, taste buds at the different sites can respond to all of the taste stimuli. For neuroenologists, this means that when tasting a wine, it should be swished over the whole surface of the tongue to stimulate the taste buds in all of the papillae.

It is important to remember that taste buds are not confined to the tongue; they are also present on the palate, at the back of the roof of the mouth, on the tonsils, and as far down as the epiglottis. The tongue needs to move the wine to reach all of these taste cells to get the maximum taste stimulation. Recently, it has been discovered that isolated taste cells are even present along the upper gastrointestinal tract, where they may be especially significant in signaling glucose intake.

A fourth type of papilla with long, slender extensions (*filiform* papillae) is present in large numbers primarily on the tongue's dorsal surface. They make the tongue rough, which is part of the motor function that enables the tongue to move food and liquid within the mouth, as described in chapter 2.

An important complication for reaching agreement during a wine tasting is that the taste apparatus (the second meaning in box 11.1) can vary widely between different individuals. First, the number of taste buds on the human tongue can vary from 2,000 to 8,000, which by itself could give a fourfold difference in detecting different components of a wine. Second, some of these people are *supertasters*; they sense a bitter test molecule as extremely bitter, in comparison with the rest of the population. This could obviously explain a difference between perceptions of bitterness in a wine. Third, a wine always combines with the saliva in stimulating the taste buds, and as we have seen, the amount and composition of the saliva varies between individuals—even for a given individual depending on the time of day and the person's emotional and hormonal status.

Sweet, Bitter, and Umami Receptors

A taste bud contains up to 70 to 80 cells of several types. The sensory type gives rise to fine hairlike extensions that protrude into the pore of the taste bud and contain the receptor molecules specific for the different types of tastant substances. The receptor mechanisms were unknown until the tools of molecular biology became available in the 1980s, and soon, as in the other sensory systems, the receptor molecules started to be identified.

A surprising—and welcome—finding was that the receptor molecules for sweet, bitter, and umami are of the same type as those for vision and smell. Their protein chains weave back and forth across the membrane seven times to form the same barrel-like structure extending across the cell membrane as shown in figure 7.3. Within this similar pattern, the receptors for sugars and bitter alkaloids have large lengths hanging outside and different sequences of elements that enable them to respond selectively to one or the other. The umami receptor lacks the outside length and also has a different pattern of elements.

It is not known whether the sugar, alkaloid (bitter), and glutamate molecules act on the external protein or in a binding pocket. Specific differences in the receptors enable different receptor types to respond to various kinds of sensory stimuli. So what had traditionally been presumed to require very different mechanisms for vision, smell, and taste turned out to involve subtle adaptations of a single molecular type. Even more amazing, these same kinds of molecules with appropriate adaptations are involved in signaling by neurotransmitters between nerve cells. It is an excellent example of the power of molecular biology, and the strategy of nature, to build diversity on common patterns. Our understanding of the molecular basis of wine taste benefits directly from these insights.

The taste molecules from the wine in the mouth wash into the pores and activate the receptors, which react by essentially giving a microkick to an attached molecule called a *G-protein*. This activates so-called second messenger molecules and a chain of responses that are similar to those in many other cells in the body that link receptors to cell responses (figure 11.2.)

These sensory cells have connections to other cells, and together they form a local microcircuit, which activates the endings of nerve fibers to carry the impulse outputs to the brain. The output is carried by different cranial nerve fibers related to the different papillae: the trigeminal (5) nerve from fungiform papillae and the facial (7) and glossopharyngeal (9) nerves from foliate and circumvallate papillae.

The simplest way to organize the system would be to have each perception arise from specific taste buds, papillae, and nerves, but nature decided otherwise. As we noted, all taste buds respond to all tastants, with small differences over the tongue's surface. Similarly, all nerves to the brain carry all of the responses to different tastants, again with some preferences. This has given rise to the concept that taste is encoded by different patterns of impulses across all of the fibers for each tastant, a

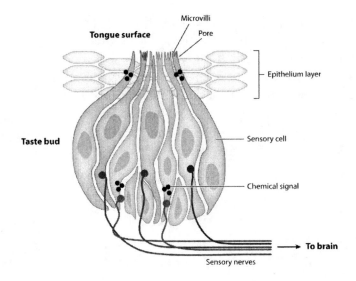

FIGURE 11.2

Taste buds consist of clusters of cells with tips giving rise to thin hairs (microvilli) carrying receptors that respond to tastants on the tongue. The cells release chemical signals (neurotransmitters) that stimulate the sensory nerves to the brain. (From J. D. Fernstrom, Non-nutritive sweeteners and obesity, *Annual Review of Food Science and Technology* 6 [2015]: 119–136)

concept called *across-fiber patterning*. This means that each tastant is represented by a different spatial pattern of input to the brain.

Be Good to Your Taste Buds

One of the remarkable properties of taste buds is that the cells turn over; that is, they die and new cells replace them. The dying cells activate a small population of stem cells within the taste bud that generate the new taste bud, on a schedule of about once a month. You may wonder why these cells die and are replaced; how does the taste bud maintain its specificity for different taste molecules matched to its nerve fibers and their connections in the brain?

We presume that the turnover reflects the fact that taste cells are exposed to the external world and that the constant effects of saliva and the actions of toxic agents that we put in our mouths stresses them.

Think of a Coca-Cola with an extreme acidic pH of 3, or a hot pepper with capsaicin, or just-boiled coffee, or a shot of 40-proof whiskey, or even a wine of 15 percent alcohol. No other sensory receptors or nerve cells are exposed to these rigors (with the exception of olfactory sensory cells exposed to the pollution of the air; as we have seen, they too are replenished through adult life). Professional wine tasters are therefore moderate in their habits. It is a reminder that for the best results in tasting wine, be good to your taste buds.

CHAPTER TWELVE

Creating Taste Perception

The taste nerves end on a cluster of cells in the brainstem called the nucleus of the solitary tract. Here, the first processing occurs to transform the taste signals from the language of the sensory world to the language of the brain world. It is also a major integrative station. We have already seen in chapter 2 that the central pattern generator for chewing is located in close proximity, where it can integrate the sensory input from the chewing muscles with the sensory input from the taste cells.

From this cluster, the cells in turn send their fibers to the thalamus, which relays them to the cerebral cortex, where they end in an area called the *insula*, a major integrative station in the frontal part of the brain. Recordings in rodents show that insula cells respond to different tastants that correlate with different taste perceptions. Some studies have obtained evidence that the taste modalities project to different sites, constituting a kind of taste map in the cortex, while others indicate that the cells responding to different modalities are intermingled; more studies are needed.

Taste in the Brain

Up until this point, we have been able to consider each of the senses as a separate system. However, we must now confront the fact that inside the brain the pathways begin to merge. As they rise higher, they increasingly merge, an essential process in forming the unified perception of flavor.

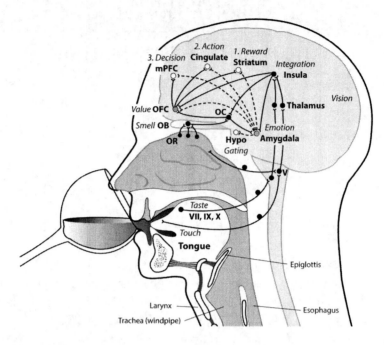

FIGURE 12.1

Simplified diagram of the taste and mouthfeel pathways to and within the cortex that contribute to conscious flavor perception. Within the cortex the two pathways converge on the same flavor system as the olfactory pathway. The actions resulting from multisensory processing are shown at the top. (Adapted from E. Rolls, Taste, olfactory, and food reward value processing in the brain, *Progress in Neurobiology* 127–128 [2015]: 64–90; and other authors)

Most of our knowledge of taste perception relevant to humans has come from neuron recordings in monkeys and brain imaging in humans. They have begun to reveal not only the taste pathway in the cerebral cortex where perception occurs but also the complex interrelations between taste, vision, olfaction, and touch. For a modern synthesis, we turn again to the approach used by Edmund Rolls at Oxford, shown in figure 12.1. The format is similar to the diagram for smell pathways in figure 9.1 and includes the mouthfeel pathway as well.

The point of this diagram for neuroenology is to indicate, first, how the brain creates the perception of taste within the context of interactions between all the sensory systems to produce the overall perception of flavor.

As in the diagram of the smell pathway, the second aim is to indicate several sequential levels of processing that take place in order to achieve this unified perception. Let us review the steps for the case of taste.

At Level 1, the brain detects only "what" the stimulus is. For a wine, this includes the memory of how it looked; its taste, especially its sweetness and acidity; its smell as an odor object; and its texture. These responses occur in the primary cortical receiving areas: $V1$ for vision; the primary taste area in the insula for taste; the OFC for smell; and the primary somatosensory area ($S1$) for texture. There may be limited interactions between the modalities, reflecting the effects we discussed previously. These responses are relatively stable, and the behavioral state of the subject does not usually significantly affect them.

These cells in turn project to the orbitofrontal cortex (OFC), which constitutes the next stage (Level 2 in figure 12.1) in processing wine flavor. Here, a proportion of the cells respond to more than one modality, for example, taste and aroma, indicating convergence of these inputs from the two primary areas. These responses are sensitive to the pleasantness of the combined stimuli and the consequent "reward" value. For wine, this presumably could vary with the wine taster's rating of the "quality" of the wine. These cells respond to the emotional and reward value through the activity of the amygdala as well as the OFC.

Finally, in Level 3, cells in areas such as the medial prefrontal cortex (mPFC) and the anterior cingulate cortex create decisions regarding relative reward values. These pathways and levels have been identified in experiments on brain mechanisms related to eating and food reward; we can hypothesize that they apply to reward values of wine tasting as well. Crucially, as we showed in chapter 9, neurons in the higher levels are subject to learning and can change their responses according to their behavioral experience. This is the neural infrastructure used to create the flavor of wine. It carries out the sensory assessment of the wine not for its food value but for its sensory and pleasantness value.

Wine tasting involves, of course, a focus on making fine distinctions. This requires the top-down control of sensory, emotional, and cognitive centers, which involves the dorsolateral prefrontal cortex in the diagram. In addition to the cortical level shown in the diagram, top-down modulation may occur at multiple levels: receptors, second messengers, synaptic relays, and cortex.

There is a very important lesson for human evolution hidden in figures 9.1 and 12.1. At the cortical level, the three levels of processing are special to the human. Rolls has discussed the evidence that rodents do not have these successive levels; in fact, they have only a tiny area equivalent to the OFC for Level 2 processing. Experiments have reported that in the rodent, sensory identification and reward evaluation are combined and occur even before reaching the cortex.

This means that the cortical processing of independent streams of sensory input at successive levels of behavioral analysis is a primate, and perhaps most highly developed human, invention. It is of adaptive value in enabling humans to carry out a more detailed analysis of food and drink flavors. Humans thereby are able to differentiate themselves to a greater degree in terms of flavor preferences. *It supports the proposal in* Neurogastronomy *that humans are more adapted for flavor perception than other species.* It also supports our hypothesis that *wine tasting takes advantage of this ability, developed for survival in selecting foods, and uses it to discriminate and enjoy the flavors of wine.*

This merging of the senses obviously raises challenges to wine tasters in teasing out the different properties of wines. We have already discussed this problem with regard to analytic versus synthetic sensory perception in chapter 9. We will have occasion to return to this theme, especially in discussing how retronasal smell contributes to wine flavor, in chapter 15.

The Taste System and Wine Tasting

We are now in a position to deal with the key question: What roles do the different taste modalities play in wine "taste"? They may be summarized as follows:

Sweet. It seems ironic that the perception of sweet is a main driver of the consumption of food and soft drinks yet is strictly limited and barely perceived in a well-balanced wine. A high level of sweetness is aversive in a normal wine but is carefully brought out and prized in a dessert wine, such as Sauternes. The absence of sweetness is called *dry*. It is a quality that comes out with aging and is much prized in better quality wines.

Bitter. Bitter would seem to have no place in giving a wine an attractive taste. However, we have noted that people can vary widely in their sensitivity to bitterness. Bitter receptors may be activated in very young red wines but are strictly limited in mature, well-balanced wines.

Sour. Mild acidity adds an "edge" to wine, giving it a liveliness; when strong it is called *tart*. Too much acid is characteristic of green fruit; too little acid makes a wine flat. The "balance" of a wine, so important in its quality, depends to a great degree on the relative amounts of sugar and acid. Acidity is also needed to support graceful aging of the wine.

Salt. Saltiness is one of the most important tastes in food and of course is essential for life, but it is seldom detectable in wines. The grape berry starts out with small amounts of salt; the finished wine maintains that low level.

Glutamate. Savoriness is important in many foods but is largely absent from wines. Levels of glutamate in the grape berry are low, as in the wine.

Not usually realized are the marked contrasts between the primary tastants in wine compared with food (figure 12.2). For food, sugar, salt, and fat are overpowering in their attractive force for human craving, as exemplified by Michael Moss's book *Salt, Sugar, Fat: How the Food Giants Hooked Us* on how the food industry manipulates these tastes and thus encourages the overeating that leads to obesity. Take soft drinks as an example. A 12-ounce canned soft drink, as already mentioned, contains 7 teaspoons of sugar. This large amount of sugar causes the brain to create an overpowering desire to drink it. Wine, on the other hand, restricts all of the tastants, together with fat, to low levels. Salt, bitter, umami, and fat in particular are to be avoided. Sugar and sour are also held to low levels, and in balance.

Given the muted taste qualities of wine in comparison with food, what is the best way to match a given wine with a given food to bring out the flavor of both in a coordinated way? This question starts with the difference between red and white wines and what foods to eat with them. Figure 12.2 provides a convenient reference for summarizing a few relevant considerations in box 12.1.

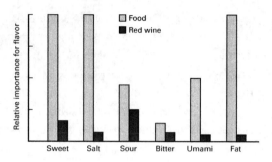

FIGURE 12.2

The reduced relative strength of different taste modalities and fat for the flavors of wine compared with the flavors of food.

BOX 12.1
Some Rules for Balancing the Taste, Aroma, and Texture of Food and Wine

White wines are usually more acid than red wines and therefore match well with fish, which often requires adding acid (lemon, etc.) to enhance flavor.

Sweetness in food increases the perception of bitterness and acidity in a wine, making it seem stronger, less sweet, and less fruity.

Foods dominated by bitter, sweet, or umami tastes accentuate any bitterness in a wine.

Salt in foods can counteract bitterness in a wine.

Acid, salty, or fatty foods suppress astringency in a wine, while sweet or spicy foods accentuate it.

Hard cheese makes wine taste softer and fuller.

Match sweet full-bodied wines with a sweet dessert.

The astringency of wine tannins reduces the viscosity of fatty foods.

Food and Wine Matching, in *The Oxford Companion to Wine*, 4th ed., ed. J. Robinson and J. Harding (Oxford: Oxford University Press, 2015), 288.

Because of the concern about the human craving for sweet foods and drinks, there are many studies characterizing how nerve cells in the brain respond to sugar stimulation. These cells are localized especially in the OFC, that part of the prefrontal cortex that houses our highest perceptual and cognitive abilities. The implication is that these cells are involved in the neural basis of the perception of sweet foods and drinks and in the emotion of pleasure and reward they give us.

For neuroenology, this poses an intriguing paradox. The brain creates a strong sense of pleasure from sugar, salt, and fat in the foods we eat. People can crave this pleasure too much, leading to cravings and negative impacts on health. Ironically, there is evidence that in rodents fed on a high-sugar diet, brain cells are activated in areas such as the insula and the OFC, which are also activated by an addiction to cocaine.

The pleasure from wine does not involve stimulation by the same tastants involved in food flavor. In addition to its "taste," wine of course contains another substance that gives pleasure: alcohol. It has been speculated that the human attraction to alcohol may have its origin in the dominant role of fruit in the diets of primates and early humans. Overripe fruit would constantly be fermenting and producing alcohol, something our ancestors would have found very enjoyable.

As always, pleasure is a two-way street, and alcohol is no exception. Too much alcohol in the short run quickly leads to drunkenness, and too much alcohol in the long run leads to addiction. Current research is revealing to what extent an excess intake of alcohol activates the same brain areas and cells that are activated by the excess intake of sugar, salt, and fat and by drugs of abuse ranging from nicotine to heroin and cocaine. It is an important reminder that the brain can create pleasure from the taste of wine only if the intake is limited to reasonable amounts. But this rule applies to life in general: both Aristotle and Confucius taught that moderation leads to happiness in all things.

CHAPTER THIRTEEN

Retronasal Smell
The Hidden Force in Wine Tasting

Until now, everything we have sensed about the wine has been conscious: the sight of the wine, its aroma as we sniff it in the glass, its touch as it enters and fills the mouth, and its taste as it stimulates the taste buds on the tongue. Now, with the wine in our mouth comes a new sense: the internal smell that is due to the retronasal pathway. Of this pathway we are *unconscious*, yet it is responsible for the major part of what we call "the taste of the wine." Being unaware of it, most wine tasters fail to appreciate this pathway. Wine professionals know about it, but their focus is usually on intensely evaluating the wine taste that they perceive, no matter how it happens. This means that retronasal smell is still poorly understood and is largely a new world in wine tasting, waiting to be explored.

We have already seen in the biomechanics of internal smell that this new world is coming into view through modern imaging methods combined with principles of dynamic airflow. Here, we will see how to extend what we know about orthonasal smell to the much more complicated activation of the brain that occurs during retronasal smell and how this provides new insights into how the brain creates the taste of wine.

Given that the orthonasal and retronasal routes both involve stimulation with smell molecules, an interesting question arises to which we have previously alluded: Are these simply two routes for stimulating the same sense, or do they actually mean that the two senses are different? If the former, the sensation of wine sniffed in when we first test the aroma should have the same qualities as the aroma sensed while breathing out. If the latter, those differences could be just quantitative, or they could be qualitative, meaning that we need to think of them as two distinct senses that can give two different kinds of clues to the wine we are drinking. It is a question researchers have been testing, and we will report the latest findings and discuss the implications for wine tasting.

The Evolution of Retronasal Smell

In all vertebrate animals, orthonasal smell is a long-distance sense, detecting molecules in the external water or, when the first fish crawled onto the land, in the air. In fish, amphibians, and reptiles, which rely on smells in their external environment, there is little separation between the internal nasal cavity and the mouth. My paleontologist colleague, Timothy Rowe of the University of Texas at Austin, has determined that the earliest mammals, emerging from reptilian-like ancestors around 250 million years ago, developed the hard palate on the roof of the mouth. This separated the nose and the mouth, enabling volatiles emitted by food at the back of the mouth to be carried into the nose when breathing out. Smell is therefore an external, long-distance sense that became adapted to being an internal, short-distance sense as well.

Fast-forwarding to a few million years ago, primates became adapted to making this internal route increasingly effective. The trend continued in our early human ancestors, especially after the invention of controlled fire. Another colleague, Richard Wrangham of Harvard University, in his book *Catching Fire: How Cooking Made Us Human*, presents strong evidence for human ancestors' adaptations to eating cooked, high-energy diets, which fed their energy-demanding brains. I have suggested that these diets also had heightened flavors of cooked meats and plants; flavors created by the brain in response to retronasal smell. Somewhere along the line, humans discovered that in addition to nutritious cooked food, fermented fruit and fruit juice were quite pleasant to taste and drink, and the love affair with fermented grapes and wine began.

A Brief History of Smell as a Dual Sense

Given this key role in human evolution and diets, when did humans first actually become aware of retronasal smell? Aristotle described the sense of smell as arising in the nose. Theophrastus (320 B.C.E.) noted a difference between sensing smells in the outside world and sensing smells arising from the mouth. A more explicit recognition came from the famous French gourmet Anthelme Brillat-Savarin. In his great book *The Physiology of Taste*, published in 1825, in "Meditation 2: On Taste," on the perception of the flavor of food and wine, he began by comparing aftertaste, "the perfume or fragrance of food," to musical harmonics:

> Without that final savouring which takes place at the back of the tongue, the whole sensation of taste would be obscure and quite incomplete.

> [On eating a peach]: It is not until the instant of swallowing, when the mouthful passes under his nasal channel, that the full aroma is revealed to him.

> In the same way, in drinking: while the wine is in the mouth, one is agreeably but not completely appreciative of it; it is not until the moment when he has finished swallowing it that a man can truly taste, consider, and discover the bouquet peculiar to each variety; and there must still be a little lapse of time before a real connoisseur can say, "It is good, or passable, or bad. By Jove, here is a Chambertin! Confound it, this is only a Suresnes!" [Chambertin dates from the twelfth century; it has been called the "King of Wines" and was Napoleon's favorite. Suresnes must have been a local wine from an area just west of Paris.]

Brillat-Savarin's observation is interesting for several reasons. It uses the word "savouring" to emphasize that, with foods or liquids in our mouths, taste as commonly applied includes the internal sense of smell. The act of swallowing is identified as producing the full contribution of internal smell to "taste"; elsewhere we have referred to it as an aroma burst. Brillat-Savarin recognized that it is the aroma of the internal smell, not the sense of taste, that is critical in discriminating wines.

And he emphasized that making these discriminations takes cognitive effort.

While the general public remained mostly oblivious to these distinctions, in the ensuing years there were occasional commentaries extending Brillat-Savarin's insights. Progress was limited by the same problem that everyone has in eating or drinking: How do we recognize the contribution of smell when it is hidden by the sense of taste? Further progress depended on well-controlled experiments separating the two.

In 1977, two experts on smell, Clair Murphy and William Cain, collaborated with an expert on taste, Linda Bartoshuk, to carry out such an experiment. Subjects were presented with different concentrations of an odorant, ethyl butyrate (smells fruity, like pineapple); a tastant, saccharin (an artificial sweetener); or a mixture of the two, to test their ability to discriminate between them. The subjects correctly identified saccharin solutions as containing only saccharin but often identified solutions containing only ethyl butyrate as a "taste." A follow-up experiment tested whether this would hold for mixtures of a smell, citral, and a tastant, sucrose, that were "congruent" because both are associated with fruit juice, and a tastant, salt, that was "incongruent." The results were the same: in both cases citral tended to be judged as a taste rather than an odor.

The authors then posed the question:

> Why should it be the rule that, since the taste and smell qualities are to be confused, smell should so commonly sacrifice its claim, so that odors are called tastes rather than vice versa? . . . This "illusion" may be mediated through the cutaneous stimulation that virtually always accompanies actual taste stimulation. . . . Neurophysiological research has only recently uncovered convergence of olfactory, gustatory, and cutaneous (trigeminal) information in the solitary nucleus. . . . The trigeminal system may serve to bind the anatomically and physiologically distinct olfactory and taste systems into a single perceptual system during eating [and wine tasting!].

This statement is fundamental to much modern research on flavor, including wine tasting. To summarize: Internal smells tend to be interpreted as "tastes." The combination is therefore an "illusion." The basis for the illusion is that the touch system localizes the food/wine to the mouth. Eventually, the systems converge in the central nervous system.

These experiments included the well-known maneuver of pinching the nose to block smell, which proved that what was interpreted as taste was in fact due to smell. In 1980, Murphy and Cain asked whether the block was complete or whether low-level stimulation could occur due to diffusion of the odor molecules from the back of the mouth. Careful testing of the subjects showed that when only the odorant was present in the mouth the odor intensity was reported to be zero. It means that the rapid transport of volatiles in the airway must take place by the flow of air, not diffusion. This flow is extremely fast, within a fraction of a second, as we discussed in chapter 4.

The study also introduced the concept of *confusion* between taste and smell. Another psychologist, Paul Rozin, picked this word up and in 1982 stimulated one of the breakthrough articles in the psychology of the senses in general and smell in particular. In "'Taste-Smell Confusions' and the Duality of the Olfactory Sense," Rozin asked: How is it that the same odor molecule "may be perceived and evaluated in two qualitatively different ways, depending on whether it is referred to the mouth or the external world"?

In attempting to understand how this confusion arises, Rozin noted that we take for granted that we see with our eyes and hear with our ears and smell with our noses, and as infants and young children, we have to learn to associate those organs with those perceptions. With flavor, it is more difficult. We have to learn to associate the internal aroma of a food or liquid with the external smell, even though interactions with taste, touch, saliva, and other factors may modify it. So it is hidden, despite the fact that it becomes a major part of the gestalt-like "odor object," a unitary percept, indissolubly merged with the other senses.

To determine the difference between perceiving an odorant by retro-nasal versus orthonasal smell, Rozin and his students did a simple test: blindfolded subjects smelled the odors of juices and soups and then tried to identify the smells as flavors while consuming them. The results: subjects had considerable difficulty identifying a specific odor when it was delivered as part of a flavor. Rozin concluded that this supported the idea that smell is a dual sense. You can try this experiment yourself and draw your own conclusion.

What could be the mechanisms of these dual senses? Rozin discussed three possibilities: There could be a gating mechanism at the receptor level, involving interactions between the tastants and the odorants in the mouth or differences in passing forward or backward within the nasal cavity. There could be interactions between the pathways in the central nervous system. And there could be a selective absorption of volatiles in the mouth that alter the actual stimulating mix of molecules. For a psychologist, Rozin concludes, the most interesting mechanisms are the more central ones in the brain. For neuroenology, they are all interesting and potentially relevant to understanding how the brain creates the taste of wine.

Importance of Retronasal Smell for Wine Tasting

Wine experts had been aware of the retronasal pathway. The excellent French book titled *Initiation à la dégustation des vins* (*Initiation to wine tasting*), published in 1978 and given to me by my mentors, contains the following: "The warming of the wine in contact with the mucus membranes of the mouth permits the least volatile substances to be perceived through the retronasal pathway."

Since these pioneering studies, the role of retronasal smell has received growing attention. However, thus far the evidence for its importance is controversial. We have noted that in the massive compendium *The Oxford Companion to Wine*, edited by Jancis Robinson and Julia Harding, aroma and flavor are closely equated, and it is stated: "The sense of smell is the most acute human tasting instrument." This presumably includes both orthonasal and retronasal smell.

Retronasal smell is being recognized as playing a dominant role in the flavors of food, so one might expect the brain to be particularly sensitive to molecules arriving in the nasal cavity via this route. However, most research on the physiology of retronasal smell has shown that the sensory thresholds for odors are higher, not lower. Leaders in this effort have been Bruce Halpern at Cornell University and Thomas Hummel in Germany. They and others concur that the ability both to detect and discriminate retronasal smell is weaker than orthonasal smell for most odorants tested. According to Ronald Jackson, the most significant contribution of retronasal olfaction in wine tasting is its involvement in the perception of the finish.

It appears that the importance of retronasal smell in wine tasting is only beginning to be studied in depth and appreciated. This may be due in large part to the difficulty of carrying out experiments in the field. The main approach thus far has been to place a tube in the mouth and blow in odorized air toward the back of the mouth so that it is carried during expiration to the nose to stimulate the olfactory receptor cells. But this does not reproduce the actual situation in which volatiles are released from the wine in the mouth. And the fact that animal experiments are usually carried out under anesthesia complicates relating the results to conscious animals and to humans.

In summary, the experience of professional wine tasters and the experiments by neuroscientists support a central role of retronasal smell in wine tasting. Further evidence comes from many directions: a consensus that retronasal smell plays a dominant role in food flavors, a belief that the sense of smell is the "most acute human tasting instrument," and adaptations of the airway to enhance retronasal smell. It seems clear that retronasal smell is a new frontier in the science and art of wine tasting and therefore should be a central focus of neuroenology.

CHAPTER FOURTEEN

Retronasal Smell
What Is So Special?

In this chapter, we flesh out the framework for understanding retronasal smell in wine perception. Box 14.1 contains the main steps. Note that they follow closely the steps for orthonasal smell outlined in chapters 7 to 9 until reaching the olfactory cortex, where the retronasal pathway begins to combine inputs from the other senses. These steps are the basis for comparing the initial wine aroma in the glass with the retronasal aroma as it contributes to the full flavor of the wine in the mouth and throat.

BOX 14.1

Steps from Retronasal Aroma Molecules to Mental Images of Aromas

1. Odor determinants differentially activate olfactory receptors.
2. An odor image of the determinants is formed in the glomerular layer.
3. An enhanced odor image is formed by glomerular layer circuits.
4. A contextual odor image is formed by lateral inhibition.
5. A content addressable memory is formed in the olfactory cortex, *modulated by input from the taste pathway.*
6. Perception occurs at the level of the orbitofrontal cortex, *combined with inputs from other sensory pathways.*
7. A mental image of the wine aroma forms, *combined with inputs from other sensory pathways.*

Testing for Retronasal Smell

We have noted that carrying out experiments on the sense of smell is challenging because we cannot see the stimuli, and it is difficult to deliver the stimuli and remove them in a closely controlled manner. These challenges are only amplified when studying retronasal smell. In human experiments, the methods are often as simple as putting saturated vapors into plastic squeeze bottles and testing the output from the nozzles. In laboratory experiments, a common method is to start with an *olfactometer*, a complicated device in which purified air is filtered through carbon cartridges to clean it, through water cartridges to humidify it, and finally through solenoid valves to control its flow. The vapor is held in bottles of odor substances and ejected through a small nozzle into the nares in the case of orthonasal smell or into the back of the mouth in the case of retronasal smell.

Odor concentration can be controlled by diluting the flow of air from a bottle or by diluting the odor concentration of the solution in the bottle. Concentrations are usually in the range of dilutions of 10^{-5} (to test for thresholds) to 10^{-1} (to test for strong stimuli), that is, from a 1 in 100,000 dilution to a 1 in 10 dilution. Few experiments have been aimed specifically at concentrations of substances pertinent to wine tasting, so we can only infer their relevance.

What Kinds of Molecules Are in Retronasal Wine Aromas?

The first question in retronasal smell is: What is the stimulus? In orthonasal smell, the molecules are unchanged as they directly enter the nose from the headspace over the wine in the glass. The different volatiles can therefore be determined by testing the aroma using sophisticated electronic instrumentation combined with expert orthonasal testing of the components, as discussed in chapter 7.

With retronasal smell, the situation is more complicated. The wine enters the mouth, and at first the volatiles are relatively unchanged. However, the wine starts mixing with the saliva. We saw in chapter 1 that this involves proteins that combine with the tannins in the wine to lend astringency to the perception of the wine taste. In addition, salivary enzymes combine with and/or break down the wine molecules. This can have several effects on the release of volatiles: It can reduce their volatility. It can produce new molecules that have less stimulating ability or change the perception. Or, it can produce new molecules that have greater volatility or stimulating effects.

These questions all make the contribution of retronasal smell to wine flavor a lot more difficult to answer. It is apparent that the range of volatiles released within the mouth evolves over time, dominated by the wine itself at the start and gradually dominated by the mixture of wine and saliva. Add to this the fact that the composition of saliva varies during the day and in different individuals. For all these reasons, the aroma we sense from the wine in our mouth gradually changes from the aroma we sniffed from the wine in the glass.

In summary, the actual stimulus in retronasal smell is a complex outcome of many variables. The role of retronasal smell in wine tasting therefore cannot be adequately studied simply by piping a wine aroma into the back of the mouth and exhaling it through the nose. A challenge to neuroenology is to develop new ways of producing, delivering, and recording the aromas of evolving mixtures of wine and saliva to get a much clearer idea of the actual stimulus during retronasal smell.

Retronasal Versus Orthonasal Smell Sensitivity

Numerous experiments measuring the sensitivity thresholds have indicated that the two flow directions do create differences in perception. As noted above, in experiments on both laboratory animals and humans, when smell stimuli are delivered by the orthonasal and retronasal routes, a higher threshold (less sensitivity) for retronasal odor perception has been reported. This seems counterintuitive because the source of the molecules in retronasal smell is the back of the mouth and the throat, so close to the nasal cavity. One possible explanation is that the concentration of volatiles from the mouth and throat is relatively higher than the concentration of smell molecules coming from the outside, so the sensitivity is adjusted to avoid overwhelming the olfactory receptors in the nasal cavity. The difference may also be related to how the experiments are performed. In particular, blowing the smell molecules into the back of the mouth may not correctly simulate how food or liquid volatiles produce them in the mouth.

This is obviously an important issue for wine tasting. If we are more sensitive to orthonasal smell, it suggests that our initial testing of the aroma before sipping may be able to detect very weak aromas that are not detectable by retronasal smell when the wine is in the mouth. We noted that orthonasal testing can begin without swirling the wine, which may enable the taster to detect the most volatile types of molecules independently of less volatile molecules released by swirling. Remember, also, that wine in the mouth immediately begins to be altered by combining with the molecules of the saliva. It seems inescapable that orthonasal and retronasal testing give different shadings of perception.

Other experiments have shown better retronasal perception. B. C. Sun and Bruce Halpern at Cornell University tested subjects for the identification of pairs of odorants in different combinations of orthonasal and retronasal routes and found the retronasal route to score most consistently. One interpretation was that the subjects in the experiment were more alert in perceiving and judging the retronasal stimuli. This is potentially relevant to wine tasting; professional wine tasters are trained to be highly focused on both orthonasal and retronasal testing, perhaps paying greater attention to retronasal testing because of the greater complexity of the flavor percept.

Another possible contribution to the difference between retronasal and orthonasal olfaction is the direction of the absorbance of odor molecules across the olfactory receptor sheet. However, as noted in chapter 4, it is likely that the differential responses of different odor receptors play the major role in both routes.

All of these factors indicate that retronasal smell differs in certain respects from orthonasal smell. For wine tasting, it means that, whether or not they are separate senses, they provide slightly different access to the composition of the wine aromas. Expert wine tasters can take advantage of any differences to learn more ways to discriminate between wines.

Retronasal Smell Images

The differential responses of the olfactory receptor cells are sent to the olfactory glomeruli in the olfactory bulb to set up the patterns of response that encode the stimulating molecules in what we have called odor images. The same processing principle applies: all the cells expressing a given receptor send their fibers to the same glomerulus, a mechanism that raises the signal-to-noise ratio at that glomerulus, which enhances the response encoded by that receptor.

Ideally, we would like to know the odor images that the wine's aroma produces. It is an experiment I have longed to perform, to compare with orthonasal images and with the images obtained in our animal experiments. This level of odor images should be sensitive to even slight changes in the spatial and/or temporal properties of an odor delivered by the two routes. Unfortunately, the olfactory bulb sits too close to the underlying bone to image with even high-strength functional magnetic resonance imaging. Other brain-imaging methods such as positron emission tomography (PET) also lack the needed high resolution to see the patterns in the glomerular layer.

Animal experiments to date have focused mostly on the odor images produced by orthonasal smell. Recently, my colleagues Shri Gautam and Justus Verhagen at the Pierce Laboratory in New Haven for the first time recorded both orthonasal and retronasal responses in animal experiments. They showed that the patterns overlap, which suggests that the encoding into odor images is similar. The retronasal patterns were weaker and slower, resembling the slower and weaker responses and higher thresholds in behavioral experiments. Odor molecules with higher vapor pressures gave stronger patterns of glomerular activation. Similar results have been obtained by a team led by Verhagen and Fahmeed Hyder (Sanganahalli et al., 2016) at Yale Medical School, in which they have recorded the patterns from calcium imaging and fMRI in the same animal.

Graeme Lowe and his colleagues Yuichi Furudono and Ginny Cruz of the Monell Chemical Senses Center in Philadelphia also showed similar results: strong odor patterns overlapped considerably. However, as the stimuli grew weaker the odor patterns grew weaker and different, and retronasal responses to longer chain odor molecule patterns disappeared. The authors speculate that the weaker retronasal responses may be due to retronasal flow having less access to all the receptor cells.

These results illustrate the advantages of laboratory experiments that cannot be done in humans. However, the airflow pathways through the nasal cavity are different in the rat compared with the human, which presents challenges in comparing the results. The animal experiments do suggest that there could be subtle differences in the encoding of smells by the two routes.

From the glomerular level, the responses are transferred to the microcircuit in the olfactory bulb formed by the interactions between mitral/tufted cells and granule cell interneurons. We have seen in chapter 8 that these interactions involve lateral inhibition, which enhances the contrast between differently activated glomerular units. The mitral cells, in turn, project the outcome of these interactions to the olfactory cortex, a key step toward aroma perception.

CHAPTER FIFTEEN

Retronasal Smell

Creating the Multisensory Wine Flavor

In our discussion of orthonasal smell, figure 10.1 shows a straight through pathway for the neural basis of smell perception. But flavor is multimodal, combining smell with the other senses. To understand the aroma of wine, we need to understand when and where combining occurs. One possibility is that merging of the senses occurs only at the highest perceptual level in the neocortex. However, nature is usually more opportunistic than that and allows for different systems to begin interacting at lower processing levels. A key system for combining with smell is taste. In chapter 12, we described how the taste pathway, like the olfactory pathway, involves evermore complex interactions as it rises to create the perception of wine flavor (see figure 12.1).

How is the taste pathway related to the retronasal smell pathway? For an orientation to the discussion that follows, figure 15.1 shows the olfactory pathway as it processes retronasal smell in parallel with the taste pathway. As can be seen, the taste pathway is mainly concerned with processing taste up to the primary taste area in the insula cortex of Level 1. This area then functions in parallel with the olfactory cortex in projecting to the orbitofrontal cortex (OFC) and the amygdala in Level 2. From there, it is simplest to think of the two systems connecting in parallel to the multiple sites in Level 3. While we focus on interactions between smell and taste, we must remember that the other senses—touch, vision, and hearing—also use these connections to contribute to the multimodal perception of flavor; unfortunately, there is no room in the diagram for all of them!

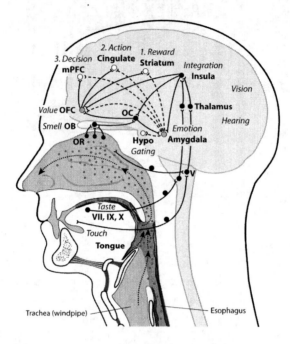

FIGURE 15.1

Simplified diagram of the multisensory wine flavor system for conscious wine tasting, combining retronasal smell and aroma burst with taste and mouthfeel. Vision and hearing connect to the same system (pathways not shown). The actions resulting from multisensory processing are shown at the top. Note the coating of wine on the tongue and walls of the throat giving rise to the aroma burst of retronasal smell. (Adapted from E. Rolls, Taste, olfactory, and food reward value processing in the brain, *Progress in Neurobiology* 127–128 [2015]: 64–90, and other authors)

It should be emphasized again that the experimental basis for this network came from studies of the aromas of food. The application to wine is a hypothesis. A casual wine taster can ignore the specific brain regions and concentrate on the fact that the perception of wine flavor is likely due to successive levels of neural processing of multiple parallel sensory inputs, with the output distributed to different behavioral actions. This provides a framework for neuroenology that enables one to relate brain imaging and brain recordings to the wine experience.

We focus on several of the critical steps in this network.

The Olfactory Cortex Begins to Integrate Smell with Taste

From the olfactory bulb, the output cells send their long axons over the surface of the olfactory cortex to make widespread connections to the pyramidal cells there, creating the "content addressable memory" of the smell stimulus, as discussed in chapter 9. The processing of odor input in the olfactory cortex uses the same mechanisms as orthonasal input. But as we have emphasized, retronasal activation is accompanied by the activation of all the other sensory pathways coming from the mouth. Since one of the main functions of the brain is to enable different systems not only to do their own thing but also to interact with other systems, we should not be surprised if this is true of the flavor system as well. And this is indeed the case.

Since the 1950s, many studies have focused on the effects of the taste system on the olfactory system and vice versa. Most of these studies have shown that tastants have a larger effect on smells than smells have on tastes. The increase is referred to as *flavor enhancement*. A 2012 study by Barry Green and his colleagues at the Pierce Laboratory in New Haven tested subjects with solutions containing odorants such as vanillin (vanilla odor), citral (lemon odor), and furaneol (strawberry odor), and tastants such as sucrose, salt, and citric acid. The subjects sampled these separately and in mixtures. The sucrose solution increased the intensity of perception of all three retronasal odors. By contrast, the odors provided only a limited enhancement of the tastes.

At about the same time, Joost Maier, Matt Wachowiak, and Donald Katz at Brandeis University and the University of Utah tested the hypothesis that this taste influence on smell could occur at the level of the olfactory cortex. Recordings from single nerve cells in the olfactory cortex of the rat showed that a significant number of cells responded selectively to taste stimuli placed on the tongue. These recordings came from the posterior part of the olfactory cortex with the greatest development of association fibers that process the input as a content addressable memory. Other studies have shown that taste input to the olfactory cortex can also come from the primary neocortical taste area in the insula. Thus, taste and smell integration is not delayed until high-level association areas in the neocortex but has already begun at lower levels.

What could be the function of this multimodal interaction? Maier and his colleagues suggested that the significance could lie in the timing sequence. Applying their hypothesis to wine, the orthonasal smell of the wine precedes the stimulation of taste during wine consumption. By contrast, during consumption, retronasal smell and taste stimulation occur much closer together. This means that the integration of taste and smell occurs in two different time sequences. For wine tasters, this may give more clues in discriminating the wine during orthonasal versus retronasal testing.

Integrating Congruent Cross-Modal Stimuli from Wine

Interactions between taste and retronasal smell are enhanced when the stimuli are "congruent"; that is, they complement each other. In 2000, this was shown by Pamela Dalton, Paul Breslin, and their colleagues. The experiment was a forced-choice detection test. The stimuli were taste plus smell (saccharin + benzaldehyde) (sweet taste + almond-cherry smell). Subthreshold amounts of saccharin were held in the mouth while the orthonasal testing of subthreshold benzaldehyde was carried out. The thresholds for almond-cherry retronasal smell were found to be decreased (made more sensitive) by 28 percent. In controls, saccharin had no odor; benzaldehyde plus water showed no effect; benzaldehyde plus MSG (noncongruent) also showed no effect.

We have emphasized that a key to wine flavor is the balance between the different sensory stimuli. The congruence between taste and smell stimuli would appear to be central to that balance. Yet we now must face the evidence from the experiment just described that the balance could involve the enhancement of otherwise undetected aromas. This suggests that quantitative measurements of wine components, such as those carried out by mass spectrometry, must always be accompanied by psychological testing of the actual flavor that the brain creates from the mixture of the components by the LCMS-O method of Acree and colleagues (see chapter 7).

Where in the brain does this enhancement due to congruence occur? Within-modality summation (enhancement between tastes, for example) is well known, presumably reflecting spatial and/or temporal summation on common targets within the same system. *Cross-modality summation* points to a central brain site or sites where nerve cells receive both inputs. We have seen that the olfactory cortex is one site. According to Dalton and her colleagues, there are several higher level sites where this might occur: first, "the *insular cortex*, which receives a convergence of inputs from visceral, taste, olfactory and somatosensory systems as well as association cortex"; second, "the *orbitofrontal cortex*, which, in monkeys, contains multisensory neurons that respond to olfactory, gustatory and visual stimuli"; and third, "the *amygdala*, which receives gustatory and olfactory input and is hypothesized to potentiate cortical information processing of stimuli that become associated through relevant life experience."

Note that these sites are part of the highest levels of cross-modal integration in the scheme of figure 15.1. Importantly, there is not just one site, but several. They constitute a distributed central "flavor substrate." In the framework of neuroenology, this is where *the brain creates the perception of flavor*. Continuing our image metaphor, it creates the central "image of flavor." In *Neurogastronomy*, this was termed the "human brain flavor system." For wine, we can call it the "human brain wine flavor system." To this core system will be added the central systems that create the extended human brain flavor system, discussed in part 3.

For the neuroenologist, these findings are very relevant. It is one thing to have taste and retronasal smell summate, or even enhance smell. However, it is quite another for taste to bring out a smell that otherwise is too weak to sense.

The binding of taste to retronasal smell, even at this early processing stage, adds to the wine taster's challenges in discriminating what is due to retronasal smell alone. But perhaps the effects are small. As we have seen, wine has limited levels of all the tastants: little salt, acid, sugar, bitter, and umami. This does not rule out taste effects on retronasal smell, and indeed the effects that basic research has revealed have parallels in the experience of wine tasters. In *Wine Tasting: A Professional Handbook*, Ronald Jackson observes that fruitiness in a wine due to aromatic compounds is increased by sugar and decreased by acidity, as would be predicted from the research results. A lower alcohol content enhances volatility (enhancing the floral notes of Riesling wine?) but may also increase the perception of acidity and astringency. This reflects the fact that wine taste is a delicate balance of cross-modal perceptual properties. Jackson notes that cross-modal effects can be unpredictable due to the varying sensitivities and experience levels of the tasters, which could explain differences in wine-tasting perceptions.

It remains to note that although we are focusing here on possible cross-modal effects in the olfactory cortex, we know from the wiring diagram (figure 8.1) that any effects in the olfactory cortex can have an impact on the olfactory bulb through the long collateral fibers from the output pyramidal cells. There is thus the possibility that any modulation of processing in the olfactory cortex also influences the input from the olfactory bulb. This is a reminder that the actions believed to happen only at one level in a sensory pathway may actually be present at lower levels.

Differential Brain Responses to Retronasal Versus Orthonasal Perception

Comparisons between orthonasal and retronasal smell have tended to focus on whether retronasal smell is as sensitive, and we have given several explanations for why they may differ. Another issue is whether the two routes might be specialized for food or nonfood odors. Dana Small and Thomas Hummel in Dresden, Germany, wished to use functional imaging to test this hypothesis. For neuroenology, the question would be whether wine odors would be sensed both orthonasally and retronasally, while nonwine odors (coming from the environment) would be sensed only orthonasally. To test this hypothesis, they delivered vapor stimuli through tubes inserted at the front and back of the nasal cavity to simulate orthonasal and retronasal stimulation, respectively. Appropriately, subjects reported that stimuli delivered to the back of the nose were perceived as coming from the mouth. Interestingly, the subjects could not tell whether stimuli were being delivered to the left or right nostril, indicating that touch cues were not being used to locate the odor source. The food odor was chocolate; the nonfood odor was lavender. Stimuli were delivered to the subject while brain scans were made to identify regions of activation of the functional imaging signals.

The most interesting result was seen with chocolate, which gave a strong and pleasant odor by both routes, while the activated brain areas differed from those due to lavender. As summarized in table 15.1, most

TABLE 15.1

Brain Areas Activated by a Food Odor But Not a Nonfood Odor

Orthonasal	Retronasal
Thalamus	
Hippocampus	
Amygdala	
Insula/operculum	Cingulate cortex-perigenual
Caudolateral orbitofrontal cortex	Medial orbitofrontal cortex
	Mouth area of cerebral cortex

Adapted from D. M. Small, J. C. Gerber, Y. E. Mak, and T. Hummel, Differential neural responses evoked by orthonasal versus retronasal odorant perception in humans, *Neuron* 47 (2005): 593–605.

of the areas activated by chocolate but not lavender were in the pathways covered in figure 15.1. The authors concluded that this argued against the idea that the two routes differ "only in the efficiency with which odors are delivered to the olfactory epithelium." They also showed that retronasal smells can sometimes be stronger than orthonasal smells. Another interesting finding was activity in the brain scans over the mouth area of the cerebral cortex for touch, presumably reflecting the association of retronasal smell with food in the mouth.

The finding of activity in the amygdala with orthonasal but not retronasal stimulation was at first puzzling. It was interpreted to indicate possibly that the amygdala response to orthonasal stimulation reflected the anticipation of the reward of the chocolate, while retronasal stimulation would occur after the food was being consumed.

The authors postulated that the results supported Paul Rozin's original suggestion that "the same olfactory stimulation may be perceived and evaluated in two qualitatively different ways depending on whether it is referred to the mouth or the external world."

Retronasal Smell and Flavor Images

In conclusion, although retronasal and orthonasal smell are similar in many respects, they are also qualitatively different in many subtle ways.

To reiterate, the perceived aroma may differ in intensity; the contribution of saliva may affect the volatiles; the direction of the airflow over the olfactory receptor cells differs; the associated activation of taste, mouthfeel, and mouth sounds pathways is specific to the retronasal route; and the distinctions between food and nonfood volatiles may be important. This means that the aromas associated with retronasal and orthonasal smell have qualitative differences that the wine taster can use to identify and enjoy.

As in orthonasal smell, the initial smell image in the olfactory pathway is changed into a mental image of smell in the cerebral cortex. By the time this has occurred in the retronasal pathway the smell image is thoroughly combined with taste, tactile, and visual stimuli (as well as auditory stimuli from the wine in the mouth) into the integrated perception of wine "taste" or flavor. The key point is that retronasal smell never occurs by itself; it always occurs with every other sensory pathway stimulated by the wine in the mouth and by the volatiles in the airway. This integration occurs at several steps along the way and by interactions with multiple sites at these steps. In this way is created one of the most complex, and rewarding, of human flavor experiences.

PART III

How Central Brain Systems Create the Pleasure of the Taste of Wine

We have covered the motor systems controlling the dynamic flow of wine and wine volatiles and the sensory systems creating the wine flavor and have seen that they form extensive neural networks within the brain. However, we are still not finished. There are additional networks contained within the brain concerned with wider and deeper properties of behavior that are essential to wine tasting. These central networks, recognized in *Neurogastronomy* as integral to flavor, constitute the "extended human brain flavor system", which is fully engaged in the wine experience.

Given the essential roles of both orthonasal and retronasal smell in wine flavor, we begin, first, by considering how this sensitivity varies with age and gender. Second, all human behavior depends on learning and memory, which are crucial to the wine experience. Third, in previous chapters we have often characterized the wine experience in terms of the language used to describe it, so here we consider that aspect in more depth. Finally, the ultimate evaluation of a wine lies in the degree of pleasure it gives us through the networks that create human pleasure.

CHAPTER SIXTEEN
Wine Tasting, Gender, and Aging

A paradox: wine tasting is one of the most rewarding food experiences we humans have, and we have seen that much of this is due to both orthonasal and retronasal smell. Yet at the same time, we believe our sense of smell to be quite inferior to that of most animals, ascribing this to a decline during evolution, when vision became more important.

Fortunately, evidence increasingly shows that the human sense of smell is much stronger than believed. In 2004, I expressed this notion in "The Human Sense of Smell: Are We Better Than We Think?" The answer was yes! The perception of wine flavor in fact can be used to confirm that claim, drawing as it does on the greater cognitive and limbic neural resources of the human brain to create this most complex of flavor experiences.

From this new perspective, it is useful to consider how prevalent the ability to experience this high level of flavor perception is in the human population. Richard Doty at the University of Pennsylvania has been a leading investigator of human smell abilities for many years. He introduced and has widely popularized the University of Pennsylvania Smell Identification Test (UPSIT), an effective approach to analyzing smell abilities in both normal human populations and in a variety of disorders. A question many people ask is whether the sense of smell is better in women than in men and whether it declines with age. Here, we consider those questions with regard to the abilities to detect, discriminate, and remember.

Smell Detection with Aging

Doty and Vidyulata Kamath describe the approach used in these experiments in "The Influences of Age on Olfaction: A Review." This summarizes tests of smell performance in normal populations, with the aim of establishing a baseline for comparison with the decreased smell perception that is being increasingly recognized as an early sign of neurodegenerative diseases such as Alzheimer's disease and Parkinson's disease. For this purpose, the UPSIT was used, consisting of booklets of "scratch-and sniff" odors released by scratching with a pencil tip to release microencapsulated substances appropriate to the culture of the populations being tested.

The results for odor detection are illustrated in figure 16.1. Note the early rise in children to a high level during the twenties, the maintenance of the high level until the sixties, and the decline at about age seventy. The detection of odors is slightly favored in women, becoming more evident at the older ages. The increased variability at the older ages, shown by the higher vertical bars, means that some individuals retain the sense of smell

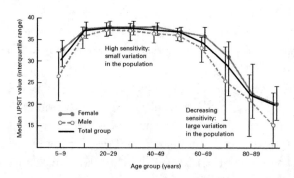

FIGURE 16.1

Test scores of orthonasal smell sensitivity using the University of Pennsylvania Smell Identification Test (UPSIT) at different ages in males and females in the United States. Note the peak plateau between ages 15 and 69, with a sharp drop-off above age 70. The numbers indicate the sizes of the sample populations, and the vertical bars indicate the variability in the populations. Note the large variability of sensitivity in the older age groups. (Modified from R. L. Doty, P. Shaman, S. L. Applebaum, R. Giberson, L Siksorski, and L. Rosenberg, Smell identification ability: changes with age, *Science* 226 [1984]: 1441–1443) as they age much better than others, who may have little sense of smell in their seventies and beyond.

These results parallel similar declines in other senses with age. For example, many older people lose the ability to hear higher frequency sounds. Fortunately, most speech occurs at lower frequencies, so conversation in person and over the phone is retained, though at reduced volume. With regard to wine, the decline in the ability to sense aromas, both by orthonasal and retronasal smell, obviously reduces older individuals' abilities to detect the more subtle components of the aromas but leaves many essential properties intact.

Smell Discrimination and Memory with Aging

In summary, odor sensitivity is highest between the ages of around twenty to seventy, and women have higher sensitivities than men. Doty and his colleagues Eric Choudhury and Paul Moberg decided to follow up by testing whether odor memory follows the same time course with age.

Their approach, reported in "Influences of Age and Sex on a Micro-encapsulated Odor Memory Test," shows how clever you have to be in these kinds of behavioral experiments. Twelve odor substances were prepared in microcapsules on scratch pads. Each participant smelled one of the odors and after a delay of 10, 30, or 60 seconds was required to identify the odor among a series of four different odors presented at 5-second intervals. The hypothesis was that the memory of the test odor would decline with increasing delays. The four odors were amyl acetate (banana-like), phenyl ethyl alcohol (rose), peppermint, and peanut.

One of the problems with this kind of test is that subjects may memorize semantic cues—words that help to identify an odor—or they may connect the smell with a particular emotion, thereby complicating the interpretation of whether one is testing only memory. To deal with this, during the delays the subjects were required to take their mind off the memory by counting out loud backward by threes, starting at 280!

The results are shown in figure 16.2. The highest scores between the ages of ten to forty-nine and the fall off in the fifties and sixties are similar to the findings with odor identification. The same applies to the higher scores for women compared with men. But the results did not support the

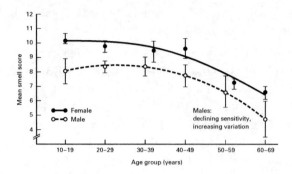

FIGURE 16.2

Test scores on a 12-item odor discrimination/memory test as a function of gender and age. Note that females have higher scores at all ages. (From E. S. Choudhury, P. Moberg, and R. L. Doty, Influences of age and sex on a microencapsulated odor memory test, *Chemical Senses* 28 [2003]: 799–805)

hypothesis: the increasing delay in the memory test had no effect in either gender or at any age. Other laboratories have reported similar results.

Choudhury, Moberg, and Doty discussed this finding as follows:

> When a subject smells an odor, there is a strong tendency to rapidly associate the odor with an object or hedonic property (i.e. to identify the source of the odor, if known) and to put into memory this association. Once this has occurred, then what is recalled later need not be the actual sensation of an odor, but simply the recollection of having smelled an odor with a name reflecting the name or source of the odor (e.g. lemon), a recollection cued by the presentation of the stimulus. Hence, often the subject is remembering, in effect, "I smelled a lemon or an odor that smells like lemon" and later, when lemon odor is presented, "I recall having smelled this stimulus, i.e. lemon." During the retention interval, knowledge of what a lemon smells like was always present in long-term memory.

This emphasizes that when we are discussing memory, is it the memory of the original stimulus or a neural or semantic cue for the stimulus? To a wine taster such as Robert Parker, it may not matter, but to a neuroenologist, it is going to the heart of how the brain creates Parker's thousands of memories.

What is the neural basis for the memory? Choudhury and his colleagues went on to note: "Wilson and Stevenson (2003) have argued that, in fact, all odors are initially encoded as 'objects' in the anterior piriform cortex, reflecting a synthesis of feature detector information from the olfactory bulb mitral cells. They hypothesize that 'odor perception is wholly dependent on the integrity of this "piriform" memory system.' "

This restates our explanation of the neural basis of smell processing in the olfactory pathway for both orthonasal and retronasal smell in chapters 9 and 15, respectively.

Male Compared with Female

The higher performance of women has invited several speculations as to its neural basis. First, it has been suggested that the odors used in these experiments may have more "salience," more behavioral relevance, in women, both at weak levels at the threshold for perception and also at higher concentrations. Women may be more familiar with the test odors and have words they immediately apply to a familiar odor that helps them to remember it. Put more simply, they may be more used to smelling different odors than men are. Second, since the odors were presented every 5 seconds, women may have less adaptation to repeated odor exposure. For neuroenology, this implies that although men have traditionally dominated the wine trade, it may be women who have the better "noses." I am reminded of this possibility when I sit down to dinner at home! And the greatest encyclopedia on wine is edited by a woman, Jancis Robinson.

Implications of Age for Wine Tasting

The decline with age in all aspects of olfactory performance (detection, identification, and memory) has also invited speculation on the mechanism. It is likely not due to one factor but many, which according to Choudhury, Moberg, and Doty could include "decreased numbers of olfactory receptor cells, decreased numbers of olfactory glomeruli, altered vascularity within the olfactory epithelium, loss of neurotrophic factors, decreased mitotic activity within the neuroepithelium, and increased viscosity of mucus." Aging is also associated with cognitive decline, which reduces the ability to use words to label odors for identification and recall.

All these factors are of obvious relevance to wine tasting. Any one of them may have an impact on the ability to detect a component of a wine aroma, identify it verbally, or remember it well enough to recall it when it occurs again. People are obviously going to vary in the impact these factors have on their appreciation of wine. Interestingly, during the middle years of peak sensitivity, there is relatively limited variation, but at the older ages, the declining average sensitivity has a widening range, meaning that the effects of aging are variable, hitting some people hard and others relatively little. This of course is true of the effects of aging on virtually all other abilities, such as vision, hearing, motor, sensory, or cognition. And, as we will see, the ability to remember plays a critical role in the wine-tasting experience.

CHAPTER SEVENTEEN

Memory and Wine Tasting

The importance of memory to smell was shown by a pioneer in the perception of smell and taste mixtures, David Laing, now at the University of New South Wales. In 1989, Laing and G. W. Francis were interested in how many components a person can identify in an odor mixture. This ability is obviously important for neuroenology and wine tasting. They trained subjects to identify seven individual odors and then tested them with various mixtures of the odors. Sound easy? Figure 17.1 shows the result. The subjects could identify each odor separately 82 percent of the time, but when present in a binary mixture with one other odor, the identification fell to 35 percent. With three odors, recall fell to only 14 percent, and with four odors, it fell to only 4 percent. When stricter accuracy was required, the percentages fell even more.

What accounts for this dramatic fall-off in the ability to remember and identify the components of a multicomponent odor? Laing and others have pursued this question in a number of studies since then, for example, testing panels of expert perfumers and flavorists. The results have been similar: a dramatic inability to identify components of increasingly complex mixtures.

In 2006, K. Marshall, Laing, and their collaborators extended this study to mixtures that contained both tastes and odors. They tested subjects with three taste stimuli (sucrose: sweet; sodium chloride: salty; and citric acid: sour) and three odor stimuli (cinnamaldehyde: cinnamon; hexenol: grassy; and pentanone: nail-polish remover), both singly and in

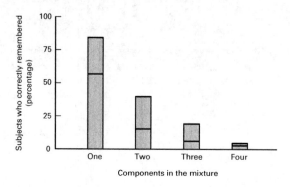

FIGURE 17.1

Classic odor memory experiment showing the decline in the percent of correct identifications of components in stimuli containing one to four odors. The top levels are for correct identification plus any errors; the lower levels are for absolute accuracy of individual components. (Modified from D. G. Laing and G. W. Francis, The capacity of humans to identify odors in mixtures, *Physiology & Behavior* 46 [1989]: 809–814)

different mixtures. The stimuli were mixed together in water, and small quantities were sipped and swallowed to ensure retronasal smell in addition to taste.

The results were very similar to the studies of odor alone: the maximum number of components that could be identified was three. The explanations for the limited ability were also similar, within either taste or smell. First, there could be competition for binding sites in the olfactory receptors or the taste receptors in the taste buds. Second, there could be suppression by lateral inhibition within the olfactory bulb or the taste centers. We have seen how important lateral inhibition is for contrast enhancement of the image, but it may be that the cost of the contrast enhancement of a single image is a limitation of the contrast enhancement of multiple images. And third, it may simply reflect a limitation in being able to hold in memory the information about the different components.

These experiments also introduced interactions between taste- and smell-processing mechanisms, so-called *intermodal interactions*. As we saw in chapter 15, these interactions may begin in the olfactory cortex and continue in the orbitofrontal cortex, where single cells combine multiple inputs from taste and smell. As we discussed, enhancement is a common response, even raising stimuli that are subthreshold by themselves to the level of conscious perception.

Memory Limits on Discrimination of Wine Aromas

Donald Wilson and Richard Stevenson have further interesting observations relevant to wine tasting. As the title of their book *Learning to Smell* indicates, the key to olfactory perception is learning:

> Single units [cells] in the rat anterior piriform [olfactory] cortex can discriminate (show minimal cross-adaptation) between binary mixtures and their components, if the mixture is familiar. With novel binary mixtures (less than 20 seconds of experience) anterior piriform cortex neurons are unable to discriminate the mixture from its components. In other words, when mixtures are familiar, cortical neurons treat them as unique objects, different from their components. . . . After experience, the cortex has learned to treat the combinations of features constituting the mixture as a unique, complete, synthetic object, distinct from the patterns of features constituting their components.

The lesson here for neuroenology is that the more experience one has in tasting wines and identifying their characteristics, the better one will learn to associate specific sets of features as synthetic objects, which collectively identify the wine being tasted. We will see that other brain mechanisms, especially language, aid in this ability.

Learning to Smell

If research suggests a limit on the number of properties we can sense in a smell, how does that square with the abilities of wine experts such as Robert Parker, who routinely reports many properties to distinguish thousands of different wines and vintages? At least part of the answer is that the properties include those due not only to wine smell but also to sensory features such as tannins, sweetness, body (due to the alcohol content), and color. We have discussed this high-level multimodal integration in the chapters on taste and retronasal smell. And finally, as Frédéric Brochet and Denis Dubordieu concluded in 2000 in a study of Parker's abilities, an important criterion is simply whether the taster likes the wine. Thus, the final evaluation of a wine taste is a combination of analysis and synthesis, of details and the gestalt—just as Wilson and Stevenson suggest. The critical role of pleasure in judging a wine is discussed in chapter 19.

We have seen that the olfactory cortex is the site of learning to discriminate between smells, an important ability for learning in wine tasting. Laboratory experiments have shown that an animal that learns to discriminate between two initial odors can learn more quickly to discriminate subsequent odors. Studies on the site of this learning have revealed cells in the piriform (olfactory) cortex that show increased excitability, firing impulses more intensely over several days in response to smell stimuli and "making associative synaptic plasticity more likely, and thus facilitating learning of subsequent odor discriminations," according to Wilson and Stevenson.

In other words, the connections between neurons in the olfactory cortex are "plastic"; they change with experience so that increased discriminating activity can lead to learned, long-lasting increases in subsequent discriminating activity. Other experiments have shown that a modulatory substance, acetylcholine, from central brain fibers involved in setting the behavioral state, is necessary for these changes to occur.

For wine tasting, this implies that learning to discriminate between two wines primes the olfactory cortex for heightened discrimination between subsequent wines, all under the modulating influence of central fibers emitting acetylcholine.

Implicit and Explicit Memory: Their Relevance to Wine Tasting

The mechanisms for plasticity in the connections between neurons are of central interest in studies of learning and memory in neuroscience. Current ideas have their origin in a proposal in 1949 by Canadian psychologist Donald Hebb in his influential book *The Organization of*

Behavior. In discussing how neural circuits can be the basis for learning and memory, he proposed, long before any experimental data existed:

> When an axon of cell A is near enough to excite cell B and repeatedly or persistently takes part in firing it, some growth process or metabolic change takes place in one or both cells such that A's efficiency, as one of the cells firing B, is increased.

The essential idea is that nerve cells have many connections between them, but the only way that two cells can be involved in learning is if one fires an impulse that induces a second cell to fire an impulse, strengthening the synapse on the other cell. If this strengthening persists, it can form the basis for a memory of that coincidence. The shorthand for this idea is, "When cells fire together, they wire together." In this way, each of the sensory circuits learns and remembers its contribution to the perception of wine flavor.

What is the mechanism at the individual synapse for strengthening or weakening? A leading theory points to excitatory synapses, such as those that commonly convey the output of one cell in a sensory pathway to the next. These synapses typically emit the neurotransmitter glutamate, which excites its target neuron by acting on two kinds of receptors, one that rapidly depolarizes the target neuron (the *AMPA receptor*); if this is sufficient, the other (the *NMDA receptor*) activates slowly and strongly. These and other mechanisms are being studied as models for Hebbian synapses. Similar mechanisms are being studied at inhibitory synapses as well.

Finally, many studies have revealed that there are different kinds of memory. Very short-term memory, often called *working memory*, is involved every time we converse with someone, holding what has just been said briefly in memory before replying and moving on to the next exchange. You use this type of memory when discussing with someone the flavor of the wine you are drinking.

Distinct from this are *longer-term* memories, which are divided into two main groups, implicit and explicit. *Implicit* is the kind of memory we use when doing something, like riding a bike. Once learned, it is innate; we can get on a bike and ride it (perhaps a bit wobbly at first) even after a long interval. By contrast, we use *explicit* memory to remember facts. This can be for specific events, called *episodic memory*, and for words, called *semantic memory*. In the case of discriminating the flavor of wine, our discussion of wine aroma in chapter 15 and figure 15.1 suggests that multiple types of memory are involved: working memory for immediately communicating our experience, implicit memory for flavors we habitually recognize in table wines we drink often, and explicit memory for new wines we have tested and have discriminated from other wines. These are hypotheses for testing by future neuroenologists!

CHAPTER EIGHTEEN

The Language of Wine Tasting

The preceding chapters have presented all the brain systems that engage in creating the flavor of wine. It is a daunting array but also an exhilarating one. We have claimed that the combination of motor, sensory, and central brain systems constitutes a flavor system in humans that is unique in the animal world. The next step in this system is our unique system of language, which enables us to describe our experience in words, both for our own use and for communicating with others.

Anyone who has tasted wine knows that describing the perception in words is one of our most difficult linguistic challenges. Much of this is due to the difficulty of describing the aromas, both orthonasal and retronasal. Some claim that this is because at the cortical level the olfactory system is distant from the cortical areas for speech and language. The area for speech reception, *Wernicke's area*, is near the auditory areas, which would seem to facilitate auditory reception and the interpretation of speech, while the area for speech formation, *Broca's area*, is near the somatosensory and motor areas that may facilitate the formation of speech. However, the visual areas at the back of the brain in the occipital lobe are as distant from these areas as the olfactory and taste areas at the front of the brain.

Patterns and Words

While it seems as if we can easily describe visual stimuli in words, this really applies only to the simplest and most familiar objects. A table can be described very precisely, as can Vincent van Gogh's painting of a sunflower. However, in *Neurogastronomy*, I pointed out that many pieces of nonrepresentational art, with very irregular shapes and the seemingly random placements of objects, by artists such as Cy Twombley and Jackson Pollock, are not easily described in words. Even more challenging is describing a face. Although we are very good at recognizing the face of someone we know, we are quite poor at describing that face in words that will allow someone else to recognize it among many similar faces. In chapter 9, we discussed Lewis Haberly's hypothesis that the processing of the irregular smell images and smell objects in the olfactory cortex is analogous to the processing of face images in higher visual association areas. From this perspective, the difficulty in using words to describe aromas and flavor is due at least in part to the complex nature of the pattern that represents them in the brain.

The Language of Wine Tasting

A comprehensive study of the language that professional wine tasters use to describe wines is needed, in order to match it with the properties of the stimuli and the brain mechanisms we have elucidated. For this purpose, it will be useful to consider Frédéric Brochet and Denis Dubordieu's careful study "Wine Descriptive Language Supports Cognitive Specificity of Chemical Senses."

Their approach fits very well with that of neuroenology in their opening statement: "The taste [in the sense of flavor] of a molecule, or of a blend of molecules, is constructed within the brain of a taster." They recognized at the outset that "taste" is likely due to the complex properties of the wine together with the experience of the individual taster and decided to focus on professional wine tasters.

They wished to answer the question: "How do wine experts proceed to describe so many fragrances in wine?" The strategy was to gather and study the descriptions of wines by four widely known wine experts,

158

TABLE 18.1

Types of Terms Used by Expert Tasters to Describe Wines

H	Hedonistic	Red	Chateau	White		
G	Old	Texture	Nice	Gold	Fresh	
F	Light	Powerful	Thin	Old	Good	Not so good
P	Good	Not so good	White			

Word categories: idealistic, odor, color, somesthesic, hedonistic

F. Brochet and D. Dubordieu, Wine descriptive language supports cognitive specificity of chemical senses, *Brain and Language* 22 (2001): 187–196.

three French and one American. From these authors' many thousands of wine descriptions, they selected a corpus of 1,000 notes from each, for a total of 4,000 tasting notes. Their method was to determine any consistency of terms within a given taster's notes for different wines and search for common terms used by all four tasters. I have attempted to summarize their results in table 18.1.

Brochet and Dubordieu comment:

> In looking at most of the word fields it is clear that they mix together visual (brown, purple), olfactory (apricot, pear), taste (acidic, sweet), trigeminal (tannic, hot), hedonistic (great, good), and idealistic (honest, personality) descriptive terms which cannot all strictly be considered to be part of a tasting vocabulary.

The main result: the different professionals used a striking diversity of terms. For taster H, for example, only 34 percent of the terms were common to the others. Similar low percentages applied to the other tasters. Of the 4,000 terms, only two words, "dark" and "black currant," were common to three lists. Surprisingly, except for one taster, most of the terms were not even taste or olfactory in nature; the other terms included "idealistic" and "hedonistic" (emotional), unrelated to the actual sensory properties.

When questioned about the strategy used, the tasters insisted that they tasted "analytically," uninfluenced by other experts' opinions or by the color or taste when describing the aroma. With regard to the primary senses, there was usually at least one descriptor for each, but it was striking that the distinct sensory qualities of smell (they do not recognize orthonasal and retronasal), taste, sight, and touch did not emerge as universal properties for wine characterization. Instead, the clear conclusion was that

> all wine descriptive language is in fact organized around wine types which we call prototypes.... What a wine taster does in front of a wine is not an analysis of its separate sensory properties but a comparison of all the cognitive associations he or she has from the wine (color, initial, aroma, and taste [in the broadest sense of flavor]) with the impressions he or she has already experienced when tasting other wines. When the taster speaks of a specific wine describing flavors, he or she mainly uses a series of words he or she has used previously for this category of wine and is not describing the specific wine. If specific wines were described independently there would be many more word groups or in the best of cases, none at all.

The ability to recognize these prototypes by idiosyncratic criteria is what most professionals possess. For example, they cite one of the tasters, a wine critic, who used quality prototypes such as "great," "enjoy," and "amazing," whereas another taster, a wine maker, used wine-making prototypes such as "oxidized," "yeast," and "woody." Each had a reference framework within which linguistic tags could be used.

Hedonistic value. The results illustrate an old truism in sensory testing that hedonistic value is always present. This refers to the emotional value of how much the taster likes or does not like the taste/flavor. Hedonistic value is closely related to *preference.* Novices usually decide if the wine is "good" or "not so good" and leave it at that; experts try to make distinctions within each. The authors noted, with only a slight French perspective, that taster P evaluated from the viewpoint of a novice, which could explain this person's extraordinary success as a wine writer. In addition, all tasters used numerical ratings, though the final ratings were based on the descriptors.

Visual cues. The color of the wine strongly influenced the word descriptors used. The descriptors for red wines related to darker colors, whereas light colors were used to describe white wines. Color was in fact the only common category among the tasters. The authors ascribe this to "the brain's necessity to retrieve a strong correlation to the world it perceives and describes in language."

Wine language is specific to the wine taster. Brochet and Dubordieu cite a study by Harry Lawless that demonstrated that "experts were not able to recognize wines based on a description given by others, even when they were experts." This resonates very closely with our discussion in chapter 8 that aromas are represented as irregular spatial images like faces, which are hard to describe in words. However, they go on to cite the discovery of the large family of olfactory receptors, which implies that individuals likely vary significantly in their populations of different receptors. Another source of variability lies in the different ways that people learn about wine and wine aromas.

Wine language and human cognition. Finally, the language that the tasters used to describe the flavor of wine appeared to reflect the cognitive structure of the brain. First, there is the convergence of pathways for smell and taste, which as we have seen begins in the olfactory cortex and continues into the insula, amygdala, and orbitofrontal cortex. This likely makes it more difficult to attempt to identify with words the separate contributions of taste and smell to the wine flavor. This multimodal merging seems to be localized to the left cerebral hemisphere.

Brochet and Dubordieu stressed the close relation of the sense of smell to the limbic structures, including the hypothalamus, which reflects smell's relevance to forming preferences for certain foods. This recalls the point we made at the start: wine appreciation is built on the mechanisms of food consumption that were once critical to human survival. The authors conclude that

> the main cognitive concern regarding flavors is whether they are good or not. This concern is so strong that even experts cannot ignore it and it is what drives the organization of their descriptive language. In this way experts are not so different from novices.

This conclusion, bringing together cognition, flavor, and the brain, is a perfect lead-in to our next chapter: the emotion of pleasure as the final criterion for judging a wine.

CHAPTER NINETEEN

Pleasure

The Final Judge in Wine Tasting

In describing how the brain creates the perception of the flavor of wine, we have focused on the contributions of the different motor and sensory systems. However, the final outcome of perceiving the flavor is to decide if we like it and if so, why. This essentially involves how much pleasure it gives us. Pleasure, one of the basic emotions, gives meaning to life in general and to specific activities, such as wine tasting, in particular. It, too, is created by the brain. Like its opposite, pain, pleasure does not exist in any particular object: *pleasure is created by the brain* from a particular input or combination of inputs to the body that promotes survival. Neuroenology therefore needs to take seriously the neural basis of how pleasure is created.

What Is Pleasure?

From the *Oxford English Dictionary*, the word "pleasure" comes from the French *plaisir*, meaning "to please." In current English, "pleasure" is defined as: "1. *The condition of consciousness induced by the enjoyment or anticipation of what is felt or viewed as good or desirable* [italics added]. The opposite of pain. 2. In bad sense: Sensuous enjoyment as a chief object of life or end in itself [that is, a life devoted exclusively to pleasure is a wasted life—at least from a puritanical point of view!]." Our definition is definitely the first: producing pleasure from wine does involve deciding if the wine is "good" and "desirable."

What brain mechanisms create the emotion of pleasure? As we have emphasized, *pleasure is created by the brain*. We have seen that the perception of flavor arises from different motor systems manipulating the wine in our mouths and the volatiles in our noses and the sensory systems feeding into higher centers in our brains. Beyond the sensory perception, two things happen at these higher centers. First, the sense of consciousness is created so that we are aware of the flavor. And second, the sense of pleasure or nonpleasure is created so that we can decide how "good" and "desirable" the wine is. Neuroenology is concerned with identifying this system in the brain.

How the Brain Creates Pleasure

This field of study began in 1956 with an article in *Scientific American*, "Pleasure Center in the Brain," by a graduate student, James Olds, and his dissertation supervisor, Peter Milner, at McGill University in Montreal, Canada. They were working in the laboratory of Donald Hebb, who had made the pioneering proposal for how neural circuits form memories (chapter 17). Olds and Milner arranged for rats to press a lever to deliver electrical stimulation to electrodes inserted into different parts of the core of the brain in order to get a reward. In most areas, the animals showed little response, but in certain areas they pressed the lever again, and again, and many times more. To a psychologist, this meant that the stimulation produced *positive reinforcement*; it was as if the activations of these areas rewarded the rats.

Many studies showed that the activations fit the classical paradigm of behavioral conditioning. The most sensitive areas were found in central basal areas of the forebrain, in the hypothalamus and cingulate cortex, among others. Later writers have characterized this work on brain self-stimulation as "one of those rare scientific discoveries that starts a new field." The same results were found in many species, including humans, who reported feelings that included pleasure, as reviewed in my book *Creating Modern Neuroscience: The Revolutionary 1950s*.

Since that time, deep electrode stimulation has been used in many ways, in conjunction with other methods, such as pharmacology, to map sites in the brain concerned with pleasure and other emotions. As in others we have reviewed, these studies are mostly concerned with the emotions related to eating, so we will interpret their possible application to wine drinking. Here, we describe two interrelated approaches as examples of the kind of research that seems relevant to pleasure from wine tasting.

One series of studies by Kent Berridge at the University of Michigan has focused on the basic emotion of "liking" to closely analyze how an animal expresses pleasure. Are our brains hardwired for liking flavorful foods? This builds on a classical observation by Jacob Steiner, a pediatrician in Israel, who tested infants just after birth for their basic reactions to sweet, salty, sour, and bitter. The infants' reactions showed the appropriate facial expressions of smiling to sweet and disgust to bitter. These and parallel experiments in rodents demonstrated that these flavor reactions are not learned at the cortical level, because the cortex has not developed in an infant, but are hardwired into the brainstem. Our adult ways of indicating liking and not-liking that emerge during development add many nuances from learning and memories. From an evolutionary point of view, these hardwired circuits ensure that ripe-tasting, nutritive foods are favored, and bitter foods are rejected. Berridge distinguished these properties of "liking" from "wanting," which can be controlled differently (we may like certain foods but we decide when we want to consume them).

In early studies, Berridge worked with Ivan de Araujo, Morten Kringelbach, and Edmund Rolls at Oxford University to identify central regions in the brain involved in the representation of pleasant and unpleasant odors. Kringelbach has summarized much of this and other work in a review with Berridge and in his appropriately titled book *The Pleasure Center: Trust Your Animal Instincts*. This has produced a consensus, shared to a large extent with many other studies, that there is a wide network of regions that constitutes a "pleasure center." Many different behaviors requiring a person to judge whether he or she feels pleasure in that behavior can activate this network. The members of the network are summarized in box 19.1.

Arising from all this work is the hypothesis that the higher association levels of the sensory pathways merge into the human brain flavor system, which includes the *human pleasure network*, to create the emotional response to the sensory inputs. This overall system in relation to

BOX 19.1
Brain Regions That Form the Pleasure Network in the Brain

Cortex

Orbitofrontal cortex
Cingulate cortex
Insular cortex

Subcortex

Nucleus accumbens
Ventral pallidum
Amygdala

Subcerebrum

Hypothalamus
Brainstem (Periaqueductal gray, Ventral tegmental area)

Adapted from K. C. Berridge and M. L. Kringelbach, Pleasure systems in the brain, *Neuron* 86 (2015): 646–664.

food has been described in *Neurogastronomy*. Applied to neuroenology, the sensory pathways for color, touch, taste, and orthonasal and retronasal smell, activated by wine, merge with the pleasure network to create the feeling of relative pleasure with the perceived wine flavor. A simple representation of these systems in the brain is shown in figure 19.1.

This only scratches the surface of the depths of the brain mechanisms that contribute to the emotion of pleasure. A major player, for example, is the neurotransmitter dopamine. Many studies have indicated a correlation between levels of pleasure and levels of dopamine in the brain. Dopamine plays a major role in reward systems. However, the relation is complex; patients with Parkinson's disease, for example, in which dopamine in the brain is depleted, are reported to have normal sensations of pleasure. It would be interesting to know if this also applies to their pleasure in drinking wine.

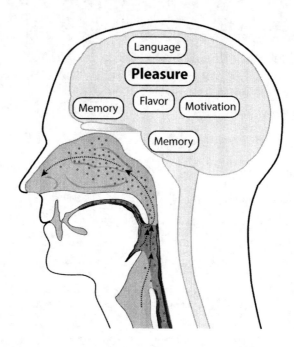

FIGURE 19.1
The combined flavor-pleasure system in the human brain.

Odor Molecule Structure and Pleasantness

Noam Sobel and his colleagues in Israel took a different approach, leading to a similar conclusion. For many years, the Sobel group has been analyzing the relation between the structure of odor molecules and the perceptions to which they give rise. The big challenge, of course, is the "high dimensionality" of the odor molecule world; the many components and shapes that the molecules can take defy finding rules for relating them to perception, as discussed in chapter 7.

Rehan Khan, Sobel, and their colleagues in an article in 2007 used a statistical approach that reduced dimensionality in both the physicochemical properties and the odor perceptions. Employing a statistical method called *principal component analysis*, they found that a large set of odor molecules could be arranged along a single primary axis, an axis similar to the primary axis of the perception of the odor molecules. They

166

were thus able to put forward the bold claim: "This allowed us to predict the pleasantness of novel molecules by their physicochemical properties alone." This focus on the single parameter of *pleasantness* is amazing in showing similar results to the use of language in describing wine properties, which also pointed to pleasantness, as described in chapter 18.

The authors caution that this combining of the perception with the odor molecule structure is completely invariant; it is not clear how universal this is when different cultures ascribe different degrees of pleasantness to the same molecules. In this context, the authors stress that the relation of molecular structure to perception is strongly dependent on learning, which can modify the relation. But it appears that olfactory pleasantness is at least partially innate, reflecting the same quality to some degree as the innate emotional responses to the different tastes.

For neuroenology, it is yet another example of how the brain attempts to integrate its different systems in creating the perception of wine. We saw this with the early combining of taste with smell responses in the olfactory pathway and the later combining of all the sensory pathways at the highest levels of integration in the cerebral cortex. Now, we can add the creation of an actual emotional response to the pleasantness of the wine experience, a pleasantness that reaches all the way back to the individual molecules in the orthonasal and retronasal smell stimuli. It is no wonder that wine flavor is a strongly synthetic perception.

Neuroenology thus provides numerous challenges to neuroscientists and wine tasters as they pursue analyzing how the brain creates the perception of flavor, as well as the pleasure created by the flavor, in this most complex of substances consumed by humans.

CHAPTER TWENTY

Practical Applications of Neuroenology to the Pleasure of Wine Tasting

If pleasure is the main criterion for wine tasters, how do wine makers produce wines to satisfy this criterion? The focus we have described on the brain is beginning to provide intriguing answers.

We now know that the flavor of a wine is created in a consumer's brain. As we noted earlier, wine preferences are due to many factors, including price, attractiveness of the label, previous experience, word of mouth, and bottle size and shape. Wine makers, for their part, try to produce wines attractive to these consumer tastes based on decisions about many factors, including the density of vines per area, harvest time, filtering, microoxygenation, and fermentation temperature. Research is ongoing in the industry on how to base these decisions on more objective criteria. The evidence reviewed in this book suggests that neuroenology can contribute to this effort. We finish by presenting several research studies that address the influence of three important factors in consumer wine choice: alcohol content, price, and expertise.

Alcohol Content and Pleasantness

An international team led by Manuel Carrerias of the Basque Center of Cognition, Brain, and Language has taken the first step in looking at how a specific wine component, the content of alcohol, affects the way the brain creates the taste of wine and therefore, consumer wine choice. This research is very much in line with the aims of neuroenology.

As reported by Ram Frost, M. Carrerias, and their colleagues, the study aimed to determine how to judge consumer preference for relative alcohol content based on brain activity. It addressed the fundamental problem that wine alcohol content has been rising globally, from the traditional 11 to 13 percent to 14 to 15 percent. Some of this rise is due to global warming, which produces higher sugar content that translates, during fermentation, into higher alcohol levels. Some is due to wine makers adding sugar during fermentation (*chaptilization*). It is commonly assumed that consumers prefer the "more powerful," "intense," and "full-bodied" flavor associated with a higher alcohol content. This is particularly true of Californian, Australian, South American, and South African wines—so-called New World wines. However, many consumers prefer the more subtle flavors of lower-alcohol "Old World" wines. This study used functional imaging to determine the brain's responses to alcohol content (of which the subjects were unaware) and its relation to conscious consumer preference.

The study used a high-alcohol-content wine (14 to 15 percent), a low-alcohol-content wine (12 to 13 percent), and a tasteless solution to monitor sites of brain activity. The hypothesis was that the high alcohol would produce stronger brain stimulation. The results showed activation in the *taste cortex* (postcentral gyrus and rolandic operculum) and the thalamic nucleus that projects to it; in the *cingulate cortex*, reflecting selective attention to taste stimuli (it links frontal and parietal attention regions with the insula and operculum taste cortex); and in the *cerebellum*, which has been shown to respond to taste intensity, presumably reflecting its role in "sensory-motor coordination during eating and drinking." Activity in the *superior temporal gyrus* may reflect its participation in a network, including the insula and thalamus, involved in taste semantics.

Unexpectedly, the low-alcohol wine produced greater stimulation of the right insula and cerebellum, regions where activity has been shown to rise with increased "taste" (in other words, flavor) intensity perception. The authors hypothesize that in the right insula this is due to the "cognitive modulation of oral sensory perception" and the "top-down modulation during attentional orienting," whereas in the cerebellum it is due to "coordinating the acquisition of sensory information." They suggest that the low-alcohol wines "induced a greater attentional orienting and exploration of the sensory attributes of wines," implying that the higher-alcohol wines overloaded the senses more. These areas can be tracked in figures 12.1 and 15.1.

In summary, their results "seem to support the intuition of some professional wine experts" that lower-alcohol content wines "have a better chance to induce greater sensitivity to the overall flavor expressed by the wine." Alcohol does have a toxic effect on body tissues, so a reduction in the sensitivity of the peripheral sense organs for taste, smell, and texture would appear to be likely and deserve further study. Of course, the central effect of alcohol is inebriation, so a lower alcohol content would appear to have the added advantage of prolonging the pleasure of the flavor over the inebriation due to the alcohol.

Price and Pleasantness

Experienced pleasure (EP) is the good feeling we get when we do something. Economists use the term to characterize the reward we get from consuming a product, such as food or wine. This EP depends on *intrinsic* factors (for example, the ingredients of a wine), the *state* of the consumer (thirsty, tired, happy), and *extrinsic* factors (promises of enjoyment, peer pressure, attractive packaging, brand loyalty). Advertisers take advantage of this because they know that there are many ways to use extrinsic factors to affect perceived pleasure.

One of the most important extrinsic factors, from an economic point of view, is price. A good question for neuroenology is: What are the neural mechanisms through which price might guide preferences for wines? Antonio Rangel and his colleagues at the Division of the Humanities and Social Sciences, California Institute of Technology, and the Stanford Graduate School of Business set out to test this by asking subjects to blind sample five cabernet sauvignon wines they were told were of increasing price, from $5 to $10 to $35 to $45 to $90. Unknown to the subjects, there were only three different wines; the $5 and $45 wines were the same (wine 1), and the $10 and $90 wines were the same (wine 2). The subjects rated them on pleasantness and intensity of taste.

The behavioral results, reported by Hilke Plassman, A. Rangel, and their colleagues, showed that the higher-priced wines were judged to be more pleasant even though they were the same wine. The authors admitted they were shocked to discover the cheapest wine was the most preferred, which they interpreted to indicate that the tasters were relatively inexperienced and preferred strong flavor, as reported by Catharine Paddock. The intensities of the wines, however, did not show this price effect. In a follow- up blind test eight weeks later of the same wines without price information, the subjects reported no differences in pleasantness between the wines.

When the subjects' brains were scanned as they tasted and swallowed the wines, activity was found for wine 1 ($5) in a cluster in the *left medial orbitofrontal cortex* (OFC) and the *left ventromedial prefrontal cortex*; a cluster in the *ventromedial prefrontal cortex* and the *rostral anterior cingulate cortex*; and also in the *dorsolateral prefrontal cortex, visual cortex, middle temporal gyrus,* and *cingulate gyrus*. Similar sites were found with wine 2 ($10); in addition, there was activity in the *amygdala, lateral* parts of the *OFC, dorsolateral prefrontal cortex, inferior* and *middle temporal gyrus,* and *posterior cingulate cortex*. In all cases, more activity correlated with higher price, even for the same wine.

The most important finding was that activity increased in the *medial OFC*, an area known to be important for the pleasure of smell, taste, flavor, and other sensory experiences. Notably, none of the areas was involved in the lower taste pathway through the pons and thalamus to the primary taste area in the insula cortex. It means that secondary influences, such as price and advertising, as well as the opinions of others around the table, are not on the sensory responses themselves but on the cognitive processes of expectations and decision making. In other words, when our brain creates the taste of wine, it combines the integration of the sensory inputs with the complex top-down modulation by our central brain systems. We think the perception is coming from the wine, but it is heavily influenced by the ecology of our brains.

In conclusion, in this study neuroeconomics meets neuroenology: EP is "a learning signal that is used by the brain to guide future choices." Neuroeconomics is thus also a new discipline that digs into the brain mechanisms to understand better how we make the choices that give us the most pleasure.

Expertise and Wine Tasting

As a final case study in using our knowledge of neuroenology to en-hance the pleasure we get from wine, let us consider how the brains of experts differ from the brains of the rest of us in creating the taste of wine. We have seen that expertise in wine tasting likely involves the inte-gration of multiple sensory pathways and higher cognitive functions in the brain. Is there direct evidence for how this applies to wine tasting?

A team led by Richard Frackowiak at the Institute of Neurology at University College, London, was among the first to study the brain's re-sponse to wine stimulation. A. Castriota-Scanderberg, Frackowiak, and their colleagues used brain imaging of subjects to focus specifically on the effect of expertise, comparing responses in sommeliers with those in naïve subjects. Interestingly, few differences between the responses oc-curred during the ingestion of the wine, which they called the "taste period." However, differences did appear after the wine was swallowed, which they called the "after-taste period." They recognized that retrona-sal smell, which we have thoroughly documented in previous chapters as the period of the "aroma burst" and "finish," dominated this period.

In this after-taste period, the results indicated that in sommeliers the wine activated a network involving, especially, the *OFC* and nearby *in-sula* in the left hemisphere. They attributed this to the integration of taste and retronasal smell in these areas. Our analysis supports this in-terpretation and adds touch to the multisensory integration, as shown in our representation of the wine flavor system activated by touch and taste (see figure 12.1) and retronasal smell (see figure 15.1). In addition, they found selective activation in the *prefrontal cortex*, also shown in our diagrams, which they attributed to "high-level cognitive processes such as working memory and selection of behavioral strategies," pro-cesses and strategies not present in naïve subjects. By comparison, the ac-tivation of *primary sensory areas* and the *amygdala* in naïve subjects was taken to indicate a more emotional processing of the sensory stimuli.

The results are very much in line with our characterization of the human brain wine flavor system. Of particular interest is the clear impli-cation that retronasal smell is the prime driver of sommelier expertise in the after-taste period, supporting our emphasis in this book. Also of in-terest is the shift toward *the left OFC* and *insula* in the sommeliers, which could be taken to track the multisensory integration and cognitive learning that occurs during the development of expertise in wine tasting.

Subsequent studies have indicated that experts may show more differential responses during the tasting period as well. They have also reported the activation of cortical areas involved in memory and language, as we have noted in previous chapters. Future studies should give evermore detailed insight into how the brain creates the taste of wine in beginners as well as experts.

Does an expert derive more pleasure than an amateur from tasting a wine? The simplest answer is that, like amateurs in every kind of endeavor, the endeavor itself is the reward, and more or less of the same general human brain flavor system is activated. What is not present at the cognitive level is present at the emotional and reward level. The amateur gets pleasure from whatever level of enthusiasm or expertise he or she may possess, and for both amateur and expert, there are always new worlds of wine flavor to create and discover.

In conclusion, we see in just these few examples how brain imaging provides a valuable tool for interpreting the roles of alcohol, price, and expertise in determining the amount of pleasure we get from wine. It joins the many other tools covered in this book to show how the new field of neuroenology contributes to combining science and art in the appreciation of wine.

APPENDIX

A Wine-Tasting Tutorial with Jean-Claude Berrouet

Jean-Claude Berrouet was the chief wine maker and technical director for Chateau Pétrus in Bordeaux for more than 40 years. Beginning in the 1960s, Berrouet and the owner, Jean-Pierre Moueix, built Pétrus to be one of the world's greatest and most expensive wines. Upon retirement in 2007, Berrouet joined Daniel Baron in Twomey Cellars in Napa Valley, California. Baron characterizes Berrouet in superlatives in combining "subtlety, intensity and balance in a wine," and "his joy of living comes through in every glass." I was fortunate to be invited to a personal winetasting at the height of his career at Pétrus.

The time was March 2003. The place: Petrus headquarters, Libourne, France. We walked down a short corridor to the main testing laboratory and entered a small modern room of only about 10 by 15 feet. Sunlight flooded the room through a large south-facing window. White walls and white countertops along three sides and a sink in one counter made it feel like a laboratory—a very clean and brightly lit laboratory.

The room was bare except for an arresting sight: on one counter stood a neat row of 10 bottles of wine—some of the very best wines that Bordeaux produced, selected by someone who knew more about them than anyone in the world.

Before arriving, I had thought we might taste several wines, like a musician plays several solos. But Jean-Claude Berrouet had prepared for us—for what? A chamber group? No, more like a small orchestra. As with any orchestra, the conductor had carefully chosen each player with care to its part to play. The wines started on the left with the youngest, two half bottles of Pomerol 2002s, followed by three half bottles of 2001s (the same Pomerols plus a Saint-Émilion for comparison); then three full bottles of the same for 1998; and finally, two half bottles of 1990 Pomerols. Thus, explained Jean-Claude, we would start with the youngest wines (one year old for the 2002, two years for the 2001, five years for the 1998, and thirteen years for the 1990) and study how the wine taste emerges and evolves with aging into its full maturity

Here we will focus on the effect of aging on taste in the four La Fleur Petrus Pomerols. Comparisons with the other wines may be found at http://medicine.yale.edu/lab/shepherd/projects/neuroenology.aspx.

Changing Wine Taste Over Thirteen Years

The rules were that Jean-Claude would lead the way with each wine, pouring it and showing how to test it. However, in making our evaluations, I would give my opinion first, and we would then discuss it. Although I would be on the spot, I also got the impression that my candid opinion, as a representative consumer as well as a scientist, was of some interest as well.

The bottles had been opened but remained stoppered. I asked him about the traditional ritual of uncorking to let the wine "breathe," which I had read actually does little good. He confirmed this; the limited access of air through the top of the bottle is of little value in allowing the wine to breathe.

He brought out two glasses and started by pouring a small amount into the glasses. He tilted the glass and gently swirled the wine around until it completely covered the walls, then emptied it into the sink. We smelled the glass: only the smell of wine—no detergent! With our glasses prepared, Jean-Claude poured a half-inch or so into the glasses, and we started the sensory analysis.

Wine 1: 2002 La Fleur Petrus Pomerol

We followed the same sequence for each wine: color, bouquet, taste, longueur, and overall impression.

Color. The color, against daylight and in the tilted glass, looked to be a hue of red very similar to the other 2002, with which Jean-Claude agreed.

Bouquet. This had a nice bouquet. I tested it quite a bit, sticking my nose in the glass, inhaling, thinking about it, resting, and repeating. I thought I detected a smell that I could only describe as "sweet." I did not like this term because it is really applicable only to the sweet modality of taste, but no other word seemed fitting. This reflected the most difficult aspect of evaluating the sensations of smell: the lack of an adequate vocabulary.

"Yes," Jean-Claude said, "that is the alcohol smell of the merlot grape. These wines are predominately merlot."

He wrote: "Nez nice, simple, alcool sweet."

Taste. I took a sip, rolled it about on my tongue, thought about the sensations, and spit it out. There seemed to be more to this wine than the previous one, but I was having trouble again with finding the words to describe it.

Jean-Claude watched me. "I suggest that you take a bigger sip. Also, you should macerate it vigorously, with movements of the tongue, jaw, and cheeks, so that it fills every part of the mouth, not just over the tongue."

He demonstrated with a sip and a vigorous working of all these muscles: "Finally, in the middle of the maceration, suck in a bit of air, as Blouin indicates, to increase the volatile smells even more."

I followed his directions as best I could. It was amazing. The nearest I can describe it is to compare it with how much more one can see by taking dark glasses off when inside or how much more one can hear by taking earplugs out of one's ears. I could begin to sense a whole new palette of qualities, though again I lacked the vocabulary to describe them. With a larger sip of wine swishing about in every corner of my mouth, it was like filling a whole room with sensations, not just a spot in the middle of the floor. I could imagine that all possible taste buds, on the tongue and back into the throat, were being stimulated, and all possible aromatic molecules were being volatilized and pushed out the back of the throat and up through the nasal cavity to retronasally stimulate the smell receptors. With Jean-Claude's help, I recognized a hint of spiciness and also a faint tannin "structure." He said there was a hint of a scent of leather, which I could barely find, either because I could not identify it or because I lacked the link between that smell and that source. Even with this increased richness, I told Jean-Claude I thought there was a nice balance of sensations; no quality seemed to be too strong, too dominant.

He seemed pleased.

"Yes," he said, "that is something I work very hard to achieve: a harmonious balance of desirable qualities."

He wrote in our notes: "Un peu de poivre et de cuir. Plaisant pas d'amertume, un peu de tannins, bien équilibre mais limité. Structure mince, harmonieux . . ." (A hint of pepper and of leather. Pleasant, not bitter, a hint of tannins, well balanced but limited. Thin structure, harmonious . . .)

Longueur. As with the other 2002, the taste in the mouth lasted only a few seconds. Jean-Claude not only noted how long the attractive qualities persisted but also recorded any traces of bitterness.

"Sometimes," he said, "with the younger wines there will be a delayed slightly bitter taste."

In the notes: ". . . assez court, petite persistance aromatique" (. . . quite short, small aromatic persistence).

Summary. This wine was like the first—relatively simple, in the sense that it seemed mostly "on the surface," with not so much complexity or depth. It was a more interesting wine, though I figured this was probably due mainly to the improvement in my sensory evaluation skills under the tutelage of my mentor. It was not surprising that it would take some time and effort to develop those skills; I just hoped it would happen quickly enough to save me further embarrassment.

I was also learning that some motor skills are involved in wine tasting. Taking in the right amount of wine; appropriately moving the tongue, jaw, and cheeks to circulate the wine in the same way throughout the entire mouth cavity; moving the volatized air into the back of the nose for retronasal stimulation—all these movements needed to be carried out in the same way for each wine, accompanied by the same degree of concentration on the sensory effects. It reminded me of studies in neuroscience, which have shown that our sensation of touch is much more sensitive when our hand actively explores a surface than when the immobile hand is passively stimulated. This is called *active touch*. It appears that there is also "active taste."

Wine 2: 2001 La Fleur Petrus Pomerol

We were now getting into a rhythm; 7 to 8 minutes for each wine.

Color. Jean-Claude emphasized the color, looking for new shades of red in the body and hints of different yellows and oranges in the tilted edges. The color of this wine was not as bright as the previous one.

In the notes: "Robe moins vive que Magdelaine" (Robe less lively than Magdelaine). *Robe* is another term for the overall color in which the wine is "clothed."

Bouquet. This had a nice nose, the strongest of the wines so far. You got the impression that the bouquet was filling the space within the glass over the wine, the so-called headspace.

This of course is why, when you are served wine at home or in a restaurant, you request that the glass be filled only about a third of the way, in order to leave headspace for sampling the bouquet. When an overzealous waiter fills the glass to the brim, he or she may want to give the impression of generosity, but it only robs you of enjoying one of the main qualities of the wine.

I sensed that the bouquet had mostly berry flavors. Jean-Claude noted that the odors were particularly of ripe berries. Ripe fresh fruit is especially associated with the production of a chemical known as *furaneol*. This was, presumably, a key chemical in human evolution because fruits have been a major foodstuff in the evolution of primates over the past 50 million years and of humans in the past several million years. Furaneol signals that a fruit is ripe, contains the most sugar, and therefore gives the most energy. The grapes for wine are harvested when they have reached ripeness, to give the most sugar and therefore the best fermentation.

In the notes: "Bon nez, assez intense, fruits murs" (Good nose, quite intense, ripe fruit).

Taste. The taste of this wine was a definite improvement over the younger wines. It was similar in the qualities of berry flavors, tannins, and sweetness, but it was more full, in the sense of filling the mouth and nose. Jean-Claude characterized it as having more "body."

In the notes: compared with the 2001, "Le vin est plus plein, plus de corps, plus de volume . . ." (The wine is more full, has more body, more volume . . .)

Longueur. Paradoxically, although the immediate sensation was stronger, it did not last longer.

In the notes: "... mes moins long" (... but less long).

Wine 3: 1998 La Fleur Petrus Pomerol

Color. This had a deeper red color than the younger vintage.

Notes: "Couleur plus sombre" (Deeper color).

Bouquet. A strong, pleasant bouquet rising out of the glass.

Notes: "Très joli bouquet. Nez complexe, mais moins que Magdelaine" (Very nice bouquet. Complex nose, but less than Magdelaine).

Taste. This was an attractive mix of fruity flavors, very full, very well balanced.

Notes: "Évolué avec l'aération, plus charme, plus fruite, plus de volume" (Evolves with aeration, more charm, more fruit, more volume).

Longueur. The taste lasted a long time, though not quite so long as the Magdelaine.

Summary. To me, this seemed like a classic wine, with a subtle color; pleasant bouquet; attractive, well-balanced taste with some depth; and a pleasant aftertaste.

Wine 4: 1990 Fleur Petrus Pomerol

We moved now to the climax of the tasting, the thirteen-year-old wines.

Color. The body of the wine was a deep red, but the subtleties only came out in the tilted wine, where the red color spread out into a translucent fringe where one could also discern brown and orange. One immediately got the impression that something significant had been happening in the bottle during the years of fermentation.

Notes: "Couleur évolué brune, orange" (Color evolved brown, orange).

Bouquet. This was a powerful sensation of complex odors. One inhaled deeply and just tried to absorb it all, like being overwhelmed by a symphony or a work of art. There was just too much to take in all at once, but one was glad for every bit of it. The fruity qualities were there in abundance, and now I recognized as well the hints of nonfruity, animal qualities.

Notes: "Nez très animal, puissant, fruits très mûrs" (Nose very animal, strong, fruits very ripe).

Taste. This was the most complex taste thus far, with a mixture of fruity qualities seemingly embedded in the skeleton of tannic properties and softened by the muted merlot smoothness. As we discussed it, Jean-Claude searched for a way to describe the sensation and remarked that the "image" of this wine was larger and more complex than that of the 1998s. In view of our hypothesis that smells generate "odor images" in the brain, it was amazing to hear Jean-Claude use the term "image" to describe the sensation of the wine. It was the first time I had heard it used outside our academic discourse to describe in practical terms the nature of a sensation aroused by smell.

In the notes, he wrote: "La structure tannique disfasait, macération alcooleuse. Pas d'amertume. Fruits confits" (Distributed tannic structure, alcohol maceration. Not bitter. Fruit conserves).

Longueur. The taste seemed to go on and on.

Summary. A wine worthy of the wait, with all the depth, strength, complexity, and nuance one expects of a well-aged wine. Words literally failed one in describing it. It confirmed my feeling that one of the critical elements in this cognitive task is obviously to link up a sensation with a word, a descriptor. This must be a central element in the training of an enologist or anyone involved in sensory testing. With vision, the descriptor words are closely linked to a primary set of terms that apply specifically to the visual scene: "edges," "corners," "colors," "motion." With taste, the classical terms "sweet," "salty," "sour," and "bitter" have obvious links to their major type of stimulus (sugar, salt, acid, noxious substances). But there is no primary vocabulary for smells. Everything is analogy. So the expert must develop a set of meaningful terms borrowed from the other senses that facilitates the dissection of the complex odor space into parts that can be reliably identified. Jean-Claude had obviously done this in an extremely effective way.

It seems remarkable to me that this cognitive task of evaluating wines is so closely linked to language. In a way, one can view this as one of the highest cognitive challenges to humans: to identify and evaluate a complex sensation that is completely separate from our conscious world as determined by the other senses. As I had remarked in the Radio France broadcast, words fail us in describing a human face, and it seems they also fail us in describing a smell, such as that of a wine. In both cases, humans are good at recognizing these particular patterns but poor at finding the words to describe them.

Overall Summary

My basic judgments (fortunately) turned out to be in line with Jean-Claude's. However, he would always draw out fine points that I had missed. They involved impressions that I either did not quite have or that on returning to the wine recognized but could not specifically characterize. My interpretation was that I lacked the linkage between the sensory impression and the appropriate vocabulary. This supports my hypothesis about the importance of the cognitive element of olfactory discrimination in human evolution.

The youngest wines were generally pleasant, but even I could sense that their color tended toward simple shades of red, and the bouquets, tastes, and longueurs were limited. However, they all had a nice balance to them, which Jean-Claude appreciated and said he strove for in his production methods.

Moving from 2002 to 2001, one sensed an expansion of the wines in terms of more interesting colors, bigger and more complex bouquets and tastes, and somewhat increasing longueurs. On a couple of occasions, Jean-Claude noted hints of tastes in the mouth—a slight bitterness, licorice flavors—that hung on the tongue for a minute or so, but I neither noticed them nor could I identify their taste until he pointed them out.

But it was not until the 1998s that the complex qualities really began to emerge from the ageing that takes place in the bottle: varied and deep colors; nose-filling bouquets; mouth-filling flavors and nose-filling aromas; and longueurs that continued for tens of seconds. And for the first time, new qualities showed themselves, which Jean-Claude characterized in such terms as "mineral," "terroir," "animal," "leathery," or "furry."

All of the tendencies in going from the 2002s to the 1998s came out yet more fully with the 1990s: browns and purples in the color edges;

deep reds in the depths; nose-filling aromas; mouth-filling tastes, with the fruity, terroir, and animal components, always nicely balanced and seeming to enlarge the "image," as he put it, of the wine; and longueurs that lasted for minutes.

Final Conversation

After a sumptuous lunch, complete with piballe, we returned later in the day to Jean-Claude's office for a final discussion. I talked further about my concept of the odor image and how I wanted to relate it to the image of the wine on which he bases his perception of the wine's qualities.

I brought up the question of whether humans, both in their everyday lives and as experts like himself, are *macrosmats*, with a superior sense of smell compared with that of other animals. He felt it was more a question of training than any superior capacity one is born with. This in itself is interesting as it indicates the role of learning and higher cognitive function in wine tasting in particular and in the human sense of smell in general.

I mentioned the common figure of 10,000 that is often given for the number of different odors that humans can discriminate and asked him how many wines he can distinguish. He said that on an average working day he tests 10 to 20 wines, no more; one becomes fatigued. On this basis, he estimated he tests about 4000 to 5000 wines a year.

I brought up the question of how to test adequately for human odor thresholds, as I have discussed with Paul Laffort and others. Jean-Claude did not know what threshold he would have, either in absolute terms or relative to others. Dedicated learning and skilled cognitive abilities, of course, are of major importance to reach his level of expertise.

And with that we said good-bye. As the train sped back to Paris, I reviewed my notes on our tutorial and began to think: Could this lead to a book on the brain and how it creates the taste of wine?

Bibliography

General Orientation to Wine Tasting

Baldy, M. W. *The University Wine Course: A Wine Appreciation Text and Self Tutorial*. San Francisco: Wine Appreciation Guild, 2009.

Barnes, A. Q&A: Jean-Claude Berrouet. May 29, 2013. Wine-searcher. http://www .wine-searcher.com/m/2013/05/q-and-a-jean-claude-berrouet.

Brillat-Savarin, A. *The Physiology of Taste, or, Meditations on Transcendental Gastronomy* (1825). Translated by M. F. K. Fisher. Washington, D.C.: Counterpoint, 1999.

Gautier, D. *Initiation à la dégustation des vins (Introduction to wine tasting)*. Paris: Tolbiac, 1978.

Jackson, R. S. *Wine Tasting: A Professional Handbook*. 2nd ed. Amsterdam: Elsevier, 2009.

Peynaud, E., with the assistance of J. Blouin. *The Taste of Wine: The Art and Science of Wine Appreciation*. 2nd ed. Translated by M. Schuster. New York: Wiley, 1983.

Robinson, J., and J. Harding, eds. *The Oxford Companion to Wine*. 4th ed. Oxford: Oxford University Press, 2015.

Schuster, M. *Essential Winetasting: The Complete Practical Winetasting Course*. Rev. ed. London: Beazley, 2009.

Shepherd, G. M. Neuroenology: How the brain creates the taste of wine. *Flavour* (2015): 4:19.

Shepherd, G. M. *Neurogastronomy: How the Brain Creates Flavor and Why It Matters*. New York: Columbia University Press, 2012.

Specific Topics

Acree, T. E., J. Barnard, and D. G. Cunningham. A procedure for the sensory analysis of gas chromatographic effluents. *Food Chemistry* 14 (1984): 273–286.

Arshamian, A., and M. Larsson. Same same but different: The case of olfactory imagery. *Frontiers in Psychology* 5 (2014): 34.

Beauchamp, J., M. Scheibe, T. Hummel, and A. J. Buettner. Intranasal odorant concentrations in relation to sniff behavior. *Chemistry & Biodiversity* 11 (2014): 619–638.

Bennaroch, E. E. Brainstem respiratory control: Substrates of respiratory failure of multiple system atrophy. *Movement Disorders* 22 (2007): 155–161.

Berridge, K. C., and M. L. Kringelbach. Pleasure systems in the brain. *Neuron* 86 (2015): 646–664.

Bieger, D., and W. Neuhuber. Neural circuits and mediators regulating swallowing in the brainstem. Part 1, Oral cavity, pharynx and esophagus. May 16, 2006. *GI Motility online.* http://www.nature.com/gimo/contents/pt1/full/gimo74.html#f3.

Brochet, F. The taste of wine in consciousness. In "Wine Tasting." Special issue, *Journal International des Sciences de la Vigne et du Vin* (1999): 19–22.

Brochet, F., and D. Dubordieu. Wine descriptive language supports cognitive specificity of chemical senses. *Brain and Language* 22 (2001): 187–196.

Buck, L. B., and R. Axel. A novel multigene family may encode odorant receptors: A molecular basis for odor recognition. *Cell* 65 (1991): 175–187.

Buettner, A., A. Beer, C. Hannig, M. Settles, and P. Schieberle. Physiological and analytical studies on flavor perception dynamics as induced by the eating and swallowing process. *Food Quality and Preference* 13 (2002): 497–504.

Burdach, K. J., and R. L. Doty. The effects of mouth movements, swallowing, and spitting on retronasal odor perception. *Physiology & Behavior* 41 (1987): 353–356.

Byrne, J. H. Welcome to *Neuroscience Online*, the open-access neuroscience electronic textbook. Neuroscience online. neuroscience.uth.tmc.edu.

Castriota-Scanderbeg, A., G. E. Hagberg, A. Cerasa, G. Committeri, G. Galati, F. Patria, S. Pitzalis, C. Caltagirone, and R. Frackowiak. The appreciation of wine by sommeliers: A functional magnetic resonance study of sensory integration. *Neuroimage* 25 (2005): 570–578.

Cavarretta, F., A. Marasco, M. L. Hines, G. M. Shepherd, and M. Migliore. Glomerular and mitral-granule cell microcircuits coordinate temporal and spatial information processing in the olfactory bulb. *Society for Neuroscience Abstracts* (2015).

Chandrashekar, J., M. A. Hoon, N. J. P. Ryba, and C. S. Zuker. The receptors and cells for mammalian taste. *Nature* 444 (2006): 288–294.

Chin, S.-T., G. T. Eyres, and P. J. Marriott. Identification of potent odourants in wine and brewed coffee using gas chromatography-olfactometry and comprehensive two-dimensional gas chromatography. *Journal of Chromatography A* 1218 (2011): 7487–7498.

Choudhury, E. S., P. Moberg, and R. L. Doty. Influences of age and sex on a micro-encapsulated odor memory test. *Chemical Senses* 28 (2003): 799–805.

Cleland, T. A., and C. Linster. On-center/inhibitory-surround decorrelation via intraglomerular inhibition in the olfactory bulb glomerular layer. *Frontiers in Integrative Neuroscience* 6 (2012): 5.

Dalton, P., N. Doolittle, H. Nagata, and P. A. S. Breslin. The merging of the senses: Integration of subthreshold taste and smell. *Nature Neuroscience* 3 (2000): 431–432.

Doty, R. L., and V. Kamath. The influences of age on olfaction: A review. *Frontiers in Psychology* 5 (2014): 20.

Dougherty, P. Somatosensory systems. In *Neuroscience Online*, edited by J. H. Byrne. http://nba.uth.tmc.edu/neuroscience/s2/chapter02.html.

Ebeler, S. E., and J. H. Thorngate. Wine chemistry and flavor: Looking into the crystal glass. *Journal of Agriculture and Food Chemistry* 57 (2009): 8098–8108.

Edgar, M., C. Dawes, and D. O'Mullane, eds. *Saliva and Oral Health*. 3rd ed. London: British Dental Association, 2004.

Falcao, L. D., G. Lytra, P. Darriet, and J. C. Barbe. Identification of ethyl 2- hydroxy-4-methylpentanoate in red wines, a compound involved in blackberry aroma. *Food Chemistry* 132 (2012): 230–236.

Frost, R., I. Quinones, M. Velhuizen, J.-I. Alava, D. Small, and M. Carrerias. What can the brain teach us about winemaking? An fMRI study of alcohol level preferences. *PLoS One* (2015): 1–11.

Furudono, Y., G. Cruz, and G. Lowe. Glomerular input patterns in the mouse olfactory bulb evoked by retronasal odor stimuli. *BMC Neuroscience* 14 (2013): 45.

Gautam, S. H., and J. V. Verhagen. Retronasal odor representations in the dorsal olfactory bulb of rats. *Journal of Neuroscience* 32 (2012): 7949–7959.

Gawal, R. Secrets of the spit bucket revealed. Recognose: Aroma Reference Technologies, http://www.aromadictionary.com/articles/wineastringency_article.html.

Gawel, R., A. Oberholster, and I. L. Francis. A "Mouth-feel Wheel": Terminology for communicating the mouth-feel characteristics of red wine. *Australian Journal of Grape and Wine Research* 6 (2000): 203–207.

Gouras, P. Color vision. July 1, 2009. In *Webvision: The organization of the retina and visual system*, edited by H. Kolb, E. Fernandez, and R. Nelson. National Center for Biotechnology Information. http://www.ncbi.nlm.nih.gov/books/NBK11537/.

Gray, H. *Anatomy of the Human Body*. 26th ed. Edited by C. M. Goss. Philadelphia: Lea & Febiger, 1954.

Green, B. G., D. Nachtigal, S. Hammond, and J. Lim. Enhancement of retronasal odors by taste. *Chemical Senses* 37 (2012): 77–86.

Guerreiro, J. R., M. Frederiksen, V. E. Bochenkov, V. De Freitas, M. G. Ferreira Sales, and D. S. Sutherland. Multifunctional biosensor based on localized surface plasmon resonance for monitoring small molecule–protein interaction. *ACS Nano* 8 (2014): 7958–7967.

Hahn, I., P. W. Scherer, and M. M. Mozell. Velocity profiles measured for airflow through a large-scale model of the human nasal cavity. *Journal of Applied Physiology* 75 (1993): 2273–2287.

Hebb, D. O. *The Organization of Behavior.* New York: Wiley, 1949.

Hiiemae, K. M., and J. B. Palmer. Food transport and bolus formation during complete feeding sequences on foods of different initial consistency. *Dysphagia* 14 (1999): 31–42.

Hiiemae, K. M., J. B. Palmer, S. W. Medicis, J. Hegener, B. S. Jackson, and D. E. Lieberman. Hyoid and tongue surface movements in speaking and eating. *Archives of Oral Biology* 47 (2002): 11–27.

Jacobs, G. H., and J. Nathans. The evolution of primate color vision. *Scientific American* 300 (2009): 56–63.

Jean, A. Brain stem control of swallowing: Neuronal network and cellular mechanisms. *Physiological Reviews* 81 (2001): 929–969.

Kent, P. F., M. M. Mozell, S. L. Youngentob, and P. Yurco. Mucosal activity patterns as a basis for olfactory discrimination: Comparing behavior and optical recordings. *Brain Research* 981 (2013): 1–11.

Khan, R. M., C. H. Luk, A. Flinker, A. Aggarwal, H. Lapid, R. Haddad, and N. Sobel. Predicting odor pleasantness from odorant structure: Pleasantness as a reflection of the physical world. *Journal of Neuroscience* 37 (2007): 10015–10023.

Kolta, A., and J. P. Lund. The mammalian brainstem chewing circuitry. In *Handbook of Brain Microcircuits*, edited by S. Grillner and G. M. Shepherd, 317–325. New York: Oxford University Press, 2010.

Kosslyn, S. M. Mental images and the brain. *Cognitive Psychology* 22 (2005): 333–347.

Kringelbach, M. L. *The Pleasure Center: Trust Your Animal Instincts.* New York: Oxford University Press, 2009.

Laing, D. G., and G. W. Francis. The capacity of humans to identify odors in mixtures. *Physiology & Behavior* 46 (1989): 809–814.

Laska, M., D. Joshi, and G. M. Shepherd. Olfactory discrimination ability of CD-1 mice for aliphatic aldehydes as a function of stimulus concentration. *Journal of Comparative Physiology A: Sensory, Neural, and Behavioral Physiology* 193 (2007): 955–961.

Lawless, H. (1984). Flavor description of white wine by "expert" and non-expert wine consumers. *Journal of Food Science* 49 (1984): 120–123.

Lieberman, D. E. *The Evolution of the Human Head.* Cambridge, Mass.: Harvard University Press, 2011.

Ligtenberg, A. J. M., and E. C. I. Veerman, eds. *Saliva: Secretions and Functions.* Monographs in Oral Science, vol. 24. Basal: Karger, 2014.

Maier, J. X., M. Wachowiak, and D. B. Katz. Chemosensory convergence on primary olfactory cortex. *Journal of Neuroscience* 32 (2012): 17037–17047.

Marshall, K., D. G. Laing, A. L. Jinks, and I. Hutchinson. The capacity of humans to identify components in complex odor–taste mixtures. *Chemical Senses* 31 (2006): 539–545.

Masaoka, Y., H. Satoh, L. Akai, and I. Homma. Expiration: The moment we experience retronasal olfaction in flavor. *Neuroscience Letters* 473 (2010): 92–96.

Mazzatenta, A., M. Pokorski, A. Di Tano, M. Caccio, and C. Di Giulio. Influence of sensory stimulation on exhaled volatile organic compounds. *Advances in Experimental Medicine and Biology–Neuroscience and Respiration* 884 (2015): 75–79.

Morrot, G., F. Brochet, and D. Dubourdieu. The color of odors. *Brain and Language* 79 (2001): 300–320.

Moss, M. *Salt, Sugar, Fat: How the Food Giants Hooked Us*. New York: Random House, 2013.

Murphy, C., and W. S. Cain. Taste and olfaction: Independence vs. interaction. *Physiology & Behavior* 24 (1980): 601–605.

Murphy, C., W. S. Cain, and L. M. Bartoshuk. Mutual action of taste and olfaction. *Sensory Processes* 1 (1977): 204–211.

Neville, K. R., and L. B. Haberly. Olfactory cortex. In *The Synaptic Organization of the Brain*, 5th ed., edited by G. M. Shepherd, 415–454. New York: Oxford University Press, 2004.

Ni, R., M. H. Michalski, E. Brown, N. Doan, J. Zinter, N. T. Ouellette, and G. M. Shepherd. Optimal directional volatile transport in retronasal olfaction. *Proceedings of the National Academy of Sciences USA* 112 (2015): 14700–14704.

Olds, J. Pleasure center in the brain. *Scientific American* 195 (1956): 105–116.

Paddock, C. Pleasure experience of wine goes up with price. *Medical News Today*, January 15, 2008. http://www.medicalnewstoday.com/articles/93947.php.

Pager J., I. Giachetti, A. Holley, and J. Le Magnen. A selective control of olfactory bulb electrical activity in relation to food deprivation and satiety in rats. *Physiology and Behaviour* 9 (1972): 573–579.

Plassman, H., J. O'Doherty, B. Shiv, and A. Rangel. Marketing actions can modulate neural representations of experienced pleasantness. *Proceedings of the National Academy of Sciences USA* 105 (2008): 1050–1054.

Rall, W., G. M. Shepherd, T. S. Reese, and M. W. Brightman. Dendro-dendritic synaptic pathway for inhibition in the olfactory bulb. *Experimental Neurology* 14 (1966): 44–56.

Rolls, E. T. Taste, olfactory, and food reward value processing in the brain. *Progress in Neurobiology* 127–128 (2105): 64–90.

Rolls, E. T., H. Critchley, R. Mason, and E. A. Wakeman. Orbitofrontal cortex neurons: Role in olfactory and visual association learning. *Journal of Neurophysiology* 75 (1996): 1970–1981.

Roper, S. D., and N. Chaudhari. Taste coding and feedforward/feedback signaling in taste buds. In *Handbook of Brain Microcircuits*, edited by S. Grillner and G. M. Shepherd, 277–283. New York: Oxford University Press, 2010.

Rowe, T. B., and G. M. Shepherd. The role of ortho-retronasal olfaction in mammalian cortical evolution. *Journal of Comparative Neurology* 524 (2015): 471–495.

Rozin, P. "Taste-smell confusions" and the duality of the olfactory sense. *Perception and Psychophysics* 31 (1982): 397–401.

Sanganahalli, B. G., M. R. Rebello, P. Herman, X. Papademetris, G. M. Shepherd, J. V. Verhagen, and F. Hyder. Comparison of glomerular activity patterns by fMRI and wide-field calcium imaging: Implications for principles underlying odor mapping. *Neuroimage* 126 (2016): 208–218.

Shepherd, G. M. *Creating Modern Neuroscience: The Revolutionary 1950s.* New York: Oxford University Press, 2010.

Shepherd, G. M. The human sense of smell: Are we better than we think? *PLoS Biology* 2 (2004): e146.

Shepherd, G. M. Smell images and the flavour system in the human brain. *Nature* 444 (2006): 316–321.

Singer, M. S. Analysis of the molecular basis for octanal interactions in the expressed rat I7 olfactory receptor. *Chemical Senses* 25 (2000): 155–165.

Small, D. M., J. C. Gerber, Y. E. Mak, and T. Hummel. Differential neural responses evoked by orthonasal versus retronasal odorant perception in humans. *Neuron* 47 (2005): 593–605.

Somers, T. C., and M. E. Evans. Spectral evaluation of young red wines: Anthocyanin equilibrium, total phenolics, free and molecular SO_2, "chemical age." *Journal of the Science of Food and Agriculture* 28 (1977): 279–287.

Sun, B. C., and B. P. Halpern. Identification of air phase retronasal and orthonasal odorant pairs. *Chemical Senses* 30 (2005): 693–706.

Trastour, S., S. Melchionna, S. Mishra, D. Zwicker, D. E. Lieberman, E. Kaxiras, and M. P. Brenner. Shape of the human nasal cavity promotes retronasal smell. *Bulletin of the American Physical Society* 60 (2015).

Vandenbeuch, A., and S. C. Kinnamon. Why do taste cells generate action potentials? *Journal of Biology* 8 (2009): 42.

Williams, J., C. Stönner, J. Wicker, N. Krauter, B. Derstroff, E. Bourtsoukidis, T. Klüpfel, and S. Kramer. Cinema audiences reproducibly vary the chemical composition of air during films, by broadcasting scene specific emissions on breath. *Scientific Reports* 6 (2016): 25464.

Wilson, D. A., and R. J. Stevenson. The fundamental role of memory in odor perception. *Trends in Neurosciences* 26 (2003): 243–247.

Wilson, D. A., and R. J. Stevenson. *Learning to Smell: Olfactory Perception from Neurobiology to Behavior.* Baltimore: Johns Hopkins University Press, 2006.

Wrangham, R. *Catching Fire: How Cooking Made Us Human.* Cambridge, Mass.: Harvard University Press, 2008.

Yokoi, M., K. Mori, and S. Nakanishi. Refinement of odor molecule tuning by dendrodendritic synaptic inhibition in the olfactory bulb. *Proceedings of the National Academy of Sciences USA* 92 (1995): 3371–3375.

Zhao, H., L. Ivic, J. M. Otaki, M. Hashimoto, K. Mikoshiba, and S. Firestein. Functional expression of a mammalian odorant receptor. *Science* 279 (1998): 237–422.

Zhao, K., P. W. Scherer, S. A. Hajiloo, and P. Dalton. Effect of anatomy on human nasal air flow and odorant transport patterns: Implications for olfaction. *Chemical Senses* 29 (2004): 365–379.